Technoprecarious

Technoprecarious

Precarity Lab

First published in 2020 by Goldsmiths Press
Goldsmiths, University of London, New Cross
London SE14 6NW

Printed and bound by Versa Press
Distribution by the MIT Press
Cambridge, Massachusetts, and London, England

A CIP record for this book is available from the British Library

ISBN 978-1-912685-98-1 (pbk)
ISBN 978-1-912685-70-7 (ebk)

www.gold.ac.uk/goldsmiths-press

Contents

Acknowledgments

We would like to thank all members, writers, and kindred spirits of the Precarity Lab network, who have made the writing and book-making possible under very unique creative circumstances. We would like to thank our book sprint facilitator, Faith Bosworth of BookSprints.net, for making sure that our group left our sprint with a completed book draft. Faith, you were really the person we needed when we needed it, and deserve much credit for everything good in here. We would also like to thank the University of Michigan's Humanities Collaboratory project, and especially Peggy McCracken, Kristin Hass, and Sheri Sytsema-Geiger for believing in this project and making sure that everyone was paid and otherwise taken care of. Finally, we extend our sincere thanks to Casidy Campbell and Sarah Snyder for all their support work, and to Irina Aristarkhova and Tung Hui-Hu for their contributions to our thinking and writing. We would also like to thank sister projects with similar aims and ethos, including After Oil, Ecology of Networks, Matsutake Worlds, and the Mukurtu archive.

Finally, thank you to our long-term research collaborators, colleagues, and fellow workers across the globe, especially in the Navajo Nation, Detroit, Palestine, China, Mexico, and Indonesia, whose insights and friendship have shaped our thinking and made this work possible.

Contributors

Cassius Adair (he/him) is an independent scholar and radio producer from Virginia. He is writing a book about transgender people and the internet and editing a book of speculative fiction about the academy.

Iván Chaar López (he/him) is an Assistant Professor in American Studies and the principal investigator of the Border Tech Lab at the University of Texas, Austin. His work engages the fields of digital media studies, Latina/o studies, and STS. He is currently writing a book on the intersecting histories of unmanned aerial systems, cybernetics, and boundary-making along the US–Mexico border.

Anna Watkins Fisher (she/her) is an Assistant Professor of American Culture at the University of Michigan, Ann Arbor and a founding member of the Precarity Lab collective. Her first book, *The Play in the System: The Art of Parasitical Resistance* (2020), theorizes parasitism as an ambivalent tactic of resistance in twenty-first-century art and politics. She is also the co-editor, with Wendy Hui Kyong Chun, of *New Media, Old Media: A History and Theory Reader* (2nd edn, Routledge, 2015).

Meryem Kamil (she/her) is Assistant Professor of Film and Media Studies at the University of California, Irvine. Her work examines new media as a tool for anti-colonial Palestinian organizing.

Cindy Lin (she/her) is a PhD candidate in the School of Information at the University of Michigan, Ann Arbor and a certificate holder in the Science, Technology, and Society (STS) program. Her research and writing draw on long-term fieldwork with state science agencies and commercial services firms to examine the politics of computational labor and data architectures for peatland fire control in Indonesia.

Silvia Lindtner (she/her) is an Assistant Professor of Information, Digital Studies, and Science, Technology, and Society (STS) at the University of Michigan. She is the Associate Director of the Center for Ethics, Society, and Computing (ESC) and a founding member of Precarity Lab. Her forthcoming book *Prototype Nation: China and the Contested Promise of Innovation* (2020) unpacks in ethnographic and historical detail how a growing distrust in Western models of progress and development, including Silicon Valley and the tech industry after the financial crisis of 2007–2008, shaped the rise of the global maker movement and the vision of China as a "new frontier" of innovation.

Lisa Nakamura (she/her) is Gwendolyn Calvert Baker Collegiate Professor of American Culture at the University of Michigan, Ann Arbor. She is the inaugural Director of the new Digital Studies Institute at the University of Michigan and a founding member of the Precarity Lab collective (precaritylab.org). She is the author of four books on race, gender, and digital media and gaming.

Cengiz Salman (he/him) is a PhD candidate in the Department of American Culture (Digital Studies) at the University of Michigan, Ann Arbor. His research broadly focuses on the relationship between digital media, algorithms, unemployment, and racial capitalism. He holds a Master of Arts degree in Social Science from the University of Chicago (2013), and a Bachelor of Arts degree in Anthropology with a specialization in Muslim Studies from Michigan State University (2011). Salman is a recipient of a Fulbright IIE Award, which he used to conduct research on urban transformation projects in Turkey from 2011 to 2012.

Kalindi Vora (she/her) is Professor of Gender, Sexuality and Women's Studies at the University of California, Davis. She is author of *Life Support: Biocapital and the New History of Outsourced Labor* (2015), and, with Neda Atanasoski, *Surrogate Humanity: Race, Robots and the Politics of Technological*

Futures (2019). She has edited anthologies and published articles in journals such as *Radical Philosophy, Ethnos: Journal of Anthropology, Current Anthropology, Social Identities, The South Atlantic Quarterly, Postmodern Culture,* and *Catalyst: Feminism, Theory, Technoscience.*

Jackie Wang (she/her) is a black studies scholar, poet, multimedia artist, and an Assistant Professor of Culture and Media at the New School's Eugene Lang College. She received her PhD in African and African American Studies at Harvard University and was recently a fellow at the Radcliffe Institute for Advanced Study. She is the author of *Carceral Capitalism* (2018), a book on the racial, economic, political, legal, and technological dimensions of the US carceral state. She has also published a number of punk zines, including *On Being Hard Femme,* and a collection of dream poems titled *Tiny Spelunker of the Oneiro-Womb.*

McKenzie Wark (she/her) is the author of *A Hacker Manifesto* (2004), *The Beach Beneath the Street* (2011), *Capital is Dead* (2018), and *Reverse Cowgirl* (2020). She is Professor of Media and Culture at Eugene Lang College and of Liberal Studies at the New School for Social Research, New York City. Wark received the Thoma Foundation 2019 Arts Writing Award in Digital Art.

A Note on Precarity

This book was written just before the onset of the COVID-19 pandemic. As it goes to publication, we are still in the midst of reckoning with our changed and changing world. We mourn the loss of life and we rage at how the trail of illness and death has had disproportionate impact on communities of color. We see in this crisis the entrenchment of an already broken system run on depletion: an intensification of environmental racism, necropolitics, and surveillance. With this book, we hope to advance a conversation about how digital technologies amplify conditions of exploitation. We invite readers to become respondents, collaborators, and comrades as new forms of technoprecarity emerge.

The Precarity Effect: On the Digital Depletion Economy

Digital technologies consolidate wealth and influence. By creating and exploiting flexible labor and by shifting accountability to users, digital technologies expand insecure conditions for racial, ethnic, and sexual minorities, women, indigenous people, migrants, the poor, and peoples in the global south.

The digital is a medium of hyper-objectification.[1] It is a fantasy of hyper-efficient and fulfilling capitalism. The digital automates and abstracts governance. It is also a commodity, a product of both hyper-visiblized and invisibilized labor. The digital is a set of technologies that mediates, intensifies, abstracts, reproduces, and generalizes existing forms of domination. The digital is representation, automation, and modularity.[2] The digital is a material system of signification and meaning-making that grates against minerals, skins, and soils.

Therefore, we formulate the titular term as a type of precarity associated with digitality. Technoprecarity is the premature exposure to death and debility that working with or being subjected to digital technologies accelerates. It is the unevenly distributed yet pervasive condition that the gig economy, toxic metals, denied welfare, and biometric surveillance systems perpetuate. We use the term technoprecarity to mark a contemporary expression of long-extant forms of violence under racial capitalism; for instance, our definition intentionally references Ruth Wilson Gilmore's influential formulation of racism as "the state-sanctioned and/or extra-legal production and exploitation of group-differentiated vulnerability to premature death."[3] Clearly, racism predates the digital. Yet we feel that it is crucial at this particular historical moment to mark the particular ways that precarity operates now, and the differential ways the digital exports precarity to the vast majority of us.

Precarious is a word with its own instabilities. To *imprecate* is to beg, or to humbly request. To be precarious, in the etymological sense of the word, is to have your ability to survive subject to the whims of a sovereign figure. In modern usage, *precarious* means at risk of physical danger or collapse; risky, perilous.

To be precarious has moved in sense from instability in relation to a lord or master to instability in relation to material conditions of existence. To be precarious is to be without a home, a meal, a wage, or to be excluded from the formal economy, often by a criminal conviction. To be precarious is to be without a safe haven. It may even be to live without love, without care.

Precarity is not a metaphor. It is a real thing, felt by bodies that can't afford to be less than healthy but are sickened by toxic materials, behaviors, economies, and environments. Precarity can be a lack of work, of income, of security. Even those with work can feel it - precarity can refer to a material and psychic condition experienced by workers whose jobs are broken up into "gigs." Surviving from gig to gig can divert you from the possibility of living any other way.

The physical and emotional labor of women and people of color has always been appropriated as a work of love, never adequately compensated even as a "gig." For us, digital networks signal not novel dystopias but old paradigms of domination (the plantation, the colony, the prison, the military–industrial complex, the laboratory, and the special economic zone).

Precarity is most intensely concentrated among bodies relegated to *zones of depletion*. These are people whose zones of habitation and existence racial capitalism has subjected to long eras of resource extraction both human and natural, leaving behind toxic and depleted environments. These environments, in turn, cause harm to the bodies of those who have been subjected to such violent rule, driving them deeper into the underground, the undergig. As more move from gig to gig, others spiral further down into depletion.

And yet precarity is generalized, expanding to include even the creative class of digital producers in the *enrichment zones* of the world. Most people are living unsupported, in a deflated "cruelly optimistic" way, replaced by machines or, worse, treated as disposable by the algorithms that increasingly condition life chances.[4] In enrichment zones, precarity is celebrated as self-empowerment, creativity, lifelong learning, and self-determination – but it is also the phenomenon of renting out your car, home, and labor without any guarantee of economic stability.

We need new vocabularies for attending to the generalized production of precarity under contemporary racial capitalism. This is not the language in which the digital dream is advertised. In the promotional brochure for digital interconnectedness, these networked lives are pictured as always hyper-productive, virtuous, connective, and efficient. But it is clear to us that both these dreams and these networks themselves are broken.[5] We won't be recruited into optimizing this network. We set ourselves a different task.

Our approach to researching questions related to digital culture emphasizes a need to critically examine the way information and communication technologies can become instruments for facilitating the exercise of power and domination, particularly along axes of race, ethnicity, gender, class, and sexuality. We are not luddites, nor are we cynical about technology. Instead, we aim to extend the humanities' concern with issues of power and precarity to an inquiry into the way digital technologies mediate social life and make certain possibilities and impossibilities available to us. Perhaps we are naively hopeful that critique can help us build more productive and optimal technological systems for enriching social life.

"Surveillance capitalism,"[6] "platform capitalism,"[7] and "control society"[8] are terms that have been invoked to mark some of these economic and social transformations. We search, instead, for fresh language that allows us to describe our transnational perspective on digitality and precarity, a language that foregrounds race, gender, nation, and empire. What language, then, describes something as seemingly ineffable as digital circulation, as vast as global capital, as tangible as flesh?

Capitalism in this book can be understood in reference to processes of wealth and value accumulation that depend on the cyclical expansion and compression of labor relations.[9] Under capitalism, where formal waged work becomes the dominant means through which people secure the commodities that they need to survive, those who have never been privileged enough to participate in waged work, who work in informal economies, or who have been displaced from labor relations and forced to try to live without a job more crucially experience economic precarity.

Our use of the term capitalism therefore recognizes how race, gender, sexuality, citizenship, coloniality, and poverty differentially determine who is privileged enough to participate in formal labor relations and who is predisposed to more intense modes of appropriation, exploitation, and immiseration. Inspired by Cedric Robinson's notion of racial capitalism, we insist that identity-based forces of domination saturate and structure capitalist social relations that are often assumed to be based entirely on the labor relation.[10] Precarity is always the effect of intersecting forces of racist, sexist, colonial, capitalist, and homo- and heteronormative domination and oppression.

Language

We need to begin by describing this generalized condition of precarity that differentiates across zones of depletion and enrichment.

We are all born under surveillance, but not all of us are equally scrutinized. The digital writes over but does not replace previous forms of producing precarity. Forms of digital surveillance, measurement, and control are premised on ranked and graded precarity that previous forms of domination generated. But the digital does not just reproduce previous forms of precarity – it generalizes them.

Let's pause this declamatory language for a second. Perhaps you, dear reader, feel included in the category of the *precariat*,[11] or maybe you want to resist that inclusion. The "we" is always a problem, a floating assembly of imagining a common condition. The "we" is best thought of as blurry at the edges and hollow at the center, rather than as a stable identifier. "We" is a double movement. It stretches outwards, away from itself and towards others; it pulls inwards, chaining together elements external to it and excluding others.

The world might be imagined as becoming newly precarious, but it has long done so unevenly. The promise of the West was that precarity could be abolished in its entirety by remaking the world in the enriched image of those who claimed to have already been modern. The fantasy of modernization, economic development, and technological and scientific innovation as a virtue is haunting its creators, for it has generated, over and over, zones of depletion. The climate itself has become volatile and toxic for all, but poisoning some environments, bodies, lungs, and skins more than others.

We find evidence of so-called depletion in surprising places, even within the borders of supposedly enriched zones, and the sources of precarity are no longer easily locatable. So how do we describe geographies and peoples that are more precarious than others? And what of those that ensure their own security at the expense of others, at the increased precaritization of others? Those who used to imagine they were safe from precarity "over there" are no longer free from vulnerability. And it is always others who will imprecate to them for their daily bread: call them (us?) the West, the global north, the first world, the developed world, the empires. Those imagined to be the precarious ones: call them the south, the global south, the third world, the "underdeveloped" world. As

Ojibwe environmentalist Winona LaDuke declares, "There is no such thing as the first, second, and third worlds; there is only an exploiting world … whether its technological system is capitalist or communist … and a host world. Native peoples, who occupy more land, make up the host world."[12]

What if we began from the impossibility of separating the resource-*enriched* world from a resource-*depleted* world?

In this book we provide a more specific and historically grounded sense of a subset of precarities that goes beyond the "end of work" discourse that wants us to be afraid of robots. We situate fears for automation and digital technology against acute experiences, experimentations, and executions of labor displacement and devaluation, asking how and why we got here.

Who We Are

We came together as Precarity Lab in 2016 as part of a University of Michigan Humanities Collaboratory-funded project to investigate the proliferation and contestation of technoprecarity. We scrutinize who and what produces the digital, and at what cost. We place race and gender, said to be obsolete in the post-human fantasy of the digital, at the center of our work. These categories of difference bear the weight of their genesis in empire and modernity. Difference is still the operating system of governance that digitality has rendered increasingly automatic, compulsory, invisible, and surveillant. The black box of computation multiplies precarity while claiming objectivity.

Precarity is not abjection. Women, trans people, people of color, and migrants have always found dignity, meaning, pleasure, and self-knowledge within precarious conditions. We have so much to learn. We think with Anna Tsing's notion of precarity as a condition of life.[13] We study the "undercommons" theorized by Stefano Harney and Fred Moten as a way to improvise on the idea of the precariat by including an affective dimension of solidarity within the cracks that harbor life and keep us going despite it all.[14]

Our collective study of precarity emerges from unequal individual relationships to it. As a baseline, we have at least the security of employment as knowledge-workers, and no small amount of economic and social power. We came together as a group that has experienced living in different parts of the world, in both the enriched and depletion zones. We are all people with passports that allow us to cross borders, even if some of our racial and gendered appearances mean that we're scrutinized in the process. Some of us work in Asia or Latin America, or in far-from-enriched parts of North America. More of us are women than are not. Most of us are cisgender, but not all. And yet we all, in one way or another, feel a need to write towards changing the world, to test what is imaginable and achievable in a damaged, depleted planet. We are a group of differently situated people in solidarity, but we are not always in agreement about how best to address wealth and resource extraction facilitated by digital technologies.

We are committed to doing this work. The university has its own reasons for financing it – Precarity Lab was funded as part of an experiment by the university to make humanities research scientific, accelerated, and fundable. The university has steadily been moving towards the lab model as a way to help solve the crisis of legitimation in the humanities. It invests in a model of collaborative problem-solving and research innovation seen in the sciences. The university seeks to professionalize graduate student training and turn faculty research and mentorship into an enterprise.

We feel more and more intensely the fragility and indescribability of our worlds. And yet we can (we must) attempt to describe the conditions that make it appear as such. Our role as knowledge workers has itself become precarious, and not just to the extent that our labor becomes casualized. As forms of social-technical knowledge become more complex, more opaque, and more black-boxed, they are designed to evade understanding. Knowledge work itself holds onto the world with an ever-more tenuous grasp.

Digital technology builds on pre-existing forms of sociotechnical domination. The myth of the digital is that it embodied and generalized the free universal subject – rational, creative, business-minded. But the digital also builds on, reproduces, generalizes and makes abstract forms of precarity inherited from the laboratories of the colony, the plantation, the factory, and the prison. We understand the lab as a method, instrument, and site that can reproduce and legitimize conditions of precarity. This entails submitting our writing and collaborative process and the larger conditions that enable them to your scrutiny.

Unpacking the Lab

We have adopted the "laboratory" (in our name and practice) to account for our highly ambivalent yet deeply entangled position in relation to ongoing attempts to upgrade and entrepreneurialize the humanities and scholarship and higher education broadly. The laboratory is a place of labor, but where labor is subordinated to the task of elaboration. In the lab, there are consistent procedures, forms of regularity that produce observable difference. The lab experiments – experiments that can be tested, verified, stabilized, and can become the prototypes for new forms of organization and governance.

The *scientific* laboratory was born out of the Enlightenment, the European project of modernization and colonization. The invention of the scientific lab produced not only the belief in facts, rationality, and truth, but it also produced the belief in the moral figure of the scientist, the objective and detached observer who stands above in the "god trick," as Donna Haraway calls it.[15] The lab served in the making of modern man and the taming of nature, land, and peoples. It legitimized the exploitation of those rendered "other," those less modern and represented as "in need" of scientific intervention.

In Robert Boyle's articulation of laboratory science in the seventeenth century, the scientist's prejudices were supposedly excluded from the lab. By the mid-twentieth century, the lab was envisioned as no longer constrained behind walls,[16] and experimenters' immunity from prejudice supposedly followed them back into the laboratory of the world. Recapturing the modernist and imperial dimension of the lab as a method, cyberneticians helped reinstate the lab's governmental mode to make sense of human and nonhuman entities together, to order the domains of the sensible and the senseless, to latch onto the promise of possibility. The lab is no longer only a space apart from the world. It is the general condition for experimentation everywhere. It is a mode of governance.

In the classic analyses of Max Weber and Michel Foucault, the emphasis is on the *regularities* of forms of modern organization or power.[17] The early scientific laboratory was imagined as a restricted space outside of the regularities of these other forms of power, a space where experiments were conducted by special kinds of scientific subjects – modest witnesses recording and interpreting their data. According to this partition of the social world, the laboratory takes up problems generated outside its walls and experiments with their conditions to make new regularities – instruments that may then be used somewhere else, by someone else. But the muddied feet of actors could drag imperial debris into what only appears to be the "objective" space of the scientific laboratory.[18]

These forms of power and experimentation became an increasingly generalized condition. Think about the city as laboratory in sociology's interest in black migration to the industrialized urban centers of the northern US.[19] This experimentation took place within the context of anti-black racism, of domination that rendered categories of being human as precarious, as precariously human. The scientific lab, like the city-lab, cannot extricate itself from social conditions in which it is embedded. In its proliferation and intensification of bio/necropolitical regularities, the lab has proven to be an engine of precarity.

When we consider the colony, the plantation, the prison, and the factory as different kinds of social-technical regularities, but also as all being versions of the *lab*, what common dynamics become visible? The lab has long been the site of continual reinvestment in the project of modernization, the reproduction of the belief in science and technology as "a moral force" that operates by "creating an ethics of innovation, yield, and result"[20] and by establishing dominance and control.[21]

Labs organize labor and people, produce and mobilize knowledge, and test and develop subjects and objects in unfree conditions. The colony was the West's ideal laboratory, spinning scientific procedures, technologies, and techniques into policy and governmentality.[22]

The colony continues to be one of the most successful laboratories. Colonial rule uses techniques of governance that turn land into "zones," regional and seemingly bounded, bordered labs that render certain terrains attractive for investment by demarcating space and the people in it as exceptions. The exceptional zone manages risk for the "experimenter," because the zone is loosely regulated (lax environmental protection regulations, lack of labor laws) and offers tax reductions as incentives to investors. The lab operates on behalf of the empire-nation, or the empire-corporation, eager to compete in the global economy. People and land are its materials for experimentation.

The plantation and the factory are linked – the plantation forms a supply chain connection with the factory. It not only produces raw materials for the factory, but also acts as a laboratory for modes of control. Slavery enabled capitalism.

As Hortense Spillers notes in "Mama's Baby, Papa's Maybe," the commodification of the slave's flesh involved not only the bondage of Africans and people of African descent. It also transformed the bodies of slaves who were no longer able to work on the plantation due to injury or illness into a valuable resource for medical research, a "living laboratory" from which scientists could extract knowledge about human anatomy and physiology from persons whose lives forced labor had already depleted.[23]

The bodies of slaves themselves also importantly formed the basis of experiments with financialization in the credit economy of the Atlantic world.[24] Their uncompensated lives and labor formed the capital that early venture investors leveraged. In the antebellum Mississippi, slaves represented "a congealed form of the capital upon which the commercial development of the Valley depended ... The cords of credit and debt—of advance and obligation—that cinched the Atlantic economy together were anchored with the mutually defining values of land and slaves: without land and slaves, there was no credit, and without slaves, land itself was valueless."[25] And without the slave plantation, there was no factory full of workers.

Unlike sugar, indigo, and other commodities produced in the tropical and semitropical colonies of European empires, cotton re-ordered global production and trade networks and gave birth to both the factory and the European proletariat. With the explosion of the cotton industry, disparate regions of the globe became linked in unprecedented ways because cotton "has two labor-intensive stages - one in the fields, the other in factories";[26] 85–90 percent of the cotton produced in America was sent to Liverpool for sale, and then processed into textiles in British factories.

Undergirded by both the raw materials and by the techniques of organizing production grown in the plantation lab, the factory played a crucial role in creating the category of "free" labor through the concentration of workers. Carefully regulating workers' efforts and times while seeking to optimize their productivity, factories disciplined workers' bodies and senses. The factory was a lab for studying the production process with the goal of generating efficiencies. Similar to the plantation, it also helped shape the soil for the production of lifeworlds, from living quarters to sites of entertainment and conviviality. These formations were simultaneously the products of the project of modernity yet they also set its conditions of possibility.

Plantation and factory are two different modes of organizing the extraction of labor and the production of standard commodities through repetition. They can also be thought of as zones of experimentation that generate new regularities. The city is another such zone. We might think of the city today as a laboratory for experiments in reproducing the legitimacy of information technology. Also here, the experiment is conducted on the most precarious bodies. Such experiments, when generalized, multiply the precarity that was one of their conditions of possibility in the first place.

The laboratory is not always about the production of knowledge, or the generation of new regularities that will be more efficient, more rational, more frictionless. Sometimes the lab seems to exist for no other reason than the desire to experiment on precarious bodies. The lab does not need to have any relation to reason; it may enact power to experiment simply as power. Such enactment is its reason.

Universities too have always been labs. The close linkage between military science and university research is an open secret. In the US context, university laboratories have been essential to advancing military technology since World War II, with large numbers of research faculty funded by Department of Defense contracts.

The contemporary university – or as some of us call it, the neoliberal university – has embraced a generalized laboratory practice in the name of efficiency, underfunding and dismantling programs that do not self-evidently bring investment into the institution. Higher education has become a service provider for affluent or debt-laden communities of students.

We aim to repurpose the lab model, working in and against it as a cover for the kinds of antiracist, anticapitalist, queer and feminist work that is often devalued by the university.

We are not claiming any equivalence or universality of the experience of precarity. But we are claiming that very different institutional forms have always been experimental zones, that they borrow techniques from each other, and that the digital generalizes and accelerates this practice. For example, for-profit online universities, online high school courses, and charter schools all tend to spring up in places that are already depleted of educational infrastructures and the resources that they aim to bring.

We struggle with and work within contradictions and ambivalences that are not easily resolved. Currently, we write these words in the Banff Centre, a world-class conference center for the arts in Banff, Canada, built on First Nations lands, and underwritten by wealth gained from the natural resource extraction industry.[27]

We are generously funded by a project whose underlying goal is to rebrand the humanities as relevant to the market economy; even as we are critical of this economization of criticality, we too are complicit in the project of making the humanities anew, as marketable to donors, as a site for treating students as human capital and cultivating faculty as entrepreneurial agents and brands. We are, as the expression has it, living the contradictions.

Thinking With

We look to black and indigenous feminism for inspiration and intervention, all the while knowing that we shouldn't expect women of color to bear the burden of solving these problems along with their many other jobs. Indigenous feminist studies, for example, thinks beyond the analysis of commodity relations and capital accumulation, prioritizing instead relationality, space and place-making. Given how many people no longer have access to homes, jobs, or economic security, covens of care or relation may be our best and most attainable bet.

Indigenous studies also focuses our understanding of precarity in relation to the material world. The rootedness of the digital in precious metals and minerals, server farms, data centers, undersea cables, and stratospheric balloons shows the network is not an abstract model of relationality that includes some and excludes others, but a built spider's web of metal, plastic, and silicon with devastating effect on the environment.

We also look to our indigenous sisters to hold ourselves accountable in our own imbrications with ongoing settler-colonialisms. Jodi Byrd's reading of Choctaw novelist LeAnn Howe's "A Chaos of Angels" provides a guiding term, *haksuba,* in reckoning with historic and ongoing injustice while also building towards a just future; "Haksuba or chaos occurs when Indians and non-Indians bang their heads together in search of cross-cultural understanding," Howe tells us.[28] Byrd explains, *haksuba* "provides a foundational ethos for indigenous critical theories that emphasize the interconnectedness and grievability embodied within and among relational kinships created by histories of oppressions."[29] As Judith Butler explains that precarious life is that which is not worthy of grief, *haksuba* provides us with a method for anticolonial organizing: a chaotic, corrective, productive, and transformative mourning.

Our grief, however, goes hand-in-hand with play. We are influenced by women of color/black/indigenous feminisms in our insistence on joy and play in the face of precarity. Saidiya Hartman's writing on "the anarchy of colored girls assembled in a riotous manner" teaches us to pay attention to the social theory produced by black girls who elaborated a theory of freedom through the improvised practice of waywardness.[30] We seek to work within the spirit of waywardness in our orientation towards the university and the laboratory, a waywardness in our orientation to our own experiences of precarity.

We agree that the commodification of black feminism – as Catherine Knight Steele points out, Audre Lorde's writing is a commodity that sells organic tote-bags on Instagram – extends our over-reliance on their labor, especially their labor in creating digital life.[31] Black digital feminism warns us against simple evocations of blackness as resistance. We use these theories instead to maintain our focus on the material and lived experiences of racial and gendered expropriation, and to name them as such.

We hope that our tone translates the pleasure we took in writing with each other, coming together, and loosening the strictures of traditional academic writing. We know that we write about forces that feel totalizing, weighty, frightening, or impossible to overcome. We experience the depletion of emotional life that racial capitalism imposes. In response, we offer what we can: a writing practice, and a practice of living, that permits imperfection and pleasure. Like the syncopated rhythms of the *bomba* drum, always in conversation with the improvised movements of the *bomba* dancer, this manifesto can be read to the rhythm of your own body.[32] This manifesto need not be read in a linear fashion from front to back, following the numbered sequence of the chapters. Let the sound and cadence of the text respond to your own affective flows. We invite you to think and feel and write and play and make your own, wherever you are in the network of precarity.

The Undergig

Digital technologies enable and entrench various forms of labor exploitation. Digitality multiplies, metastasizes, and mutates exploitation, allowing for a more rapid extension of capital's *ratio* beyond circuits of production and circulation. Capitalism in the age of digital technologies forces itself into relation with spheres of life previously outside its locus of operations. But, as scholars working within the field of post-colonial studies have suggested,[1] capitalism's expansion to new territories occurs without entirely subsuming or encapsulating these frontiers within the logic peculiar to the points of origin of its operations. This is capitalism's weak force: its under-determining flexibility allows for the extension and entrenchment of its more abstract forces into new contexts. Further accumulation of social wealth depends on labor, exploitation, inequality, poverty, immiseration, debt.

Following the Global Financial Crisis of 2007–2008, there is a heightened sense of urgency in Western-centric scholarly and public media debates to make visible and intervene in the harmful consequences of the tech industry's naïve techno-utopianism and techno-solutionism. The promise of technologies to solve complex societal, economic, and political "problems" has long masked the proliferation of exploitation and inequality behind a rhetoric of "do good," progress, individual empowerment, and democratization.

Important as this rising awareness of labor exploitation is, it retains a troubling and all-too-simplistic binary view. It often goes as follows (framed in a somewhat caricatured way): "all of us" are "free laborers" in our day-to-day use of social media platforms – Facebook is the ultimate "social factory";[2] "platform capitalism" feeds off the making of intimate and personal connection; "some of us" are Uber drivers, who labor in a highly fragmented work arrangement aimed at preventing unionization and solidarity among workers; "others" might have it even worse, coerced to work in the physically strenuous and harmful conditions of Amazon warehouses or the flexibilized services of postal delivery.

But all of this labor exploitation in the "gig economy" depends on another form of exploitation, one often rendered as somehow "deeper" down, closer to the raw material or to the machine.

Cheap labor is a precondition of the gig economy, which is why we identify these workers as part of the undergig. Undergig workers perform the often invisible labor needed to create the conditions of digital life for everyone else. Electronics production extracts value from depleted zones and from factory workers, and it produces toxicity. The undergig also often overlaps with the "global south" category yet also exceeds such categorization.

The undergig is under-protected and underpaid. Its haunting invisibility is a necessary precondition for the fantasy of a smooth-functioning and fully automated digital world to come.

Operations of Capital and Experimentation

The undergig is sometimes patterned on colonial practices of experimentation and control, and is partly the result of agents creating new practices. The undergig depends on states to maintain differential conditions of operation and to police the borders between the enriched and depleted worlds, as their value chains in part depend on maintaining the differential between them.

As Sylvia Wynter has argued, colonization was crucial for experiments with over-representing or making dominant a particular Euro-American cishet male identity within the category of the human. For Wynter, colonialism allowed for an extension of this over-representation across the globe and the identification of this image of humanity with its truth, a "truth" that continues to act as a fulcrum upon which "all our present struggles with respect to race, class, gender, sexual orientation, ethnicity, struggles over the environment, global warming, severe climate change, [and] the sharply unequal distribution of the earth resources" continues to rest.[3]

The colonial practices of the old empires tended to focus on resource extraction. The colony was a zone from which to extract cheap nature, in the form of cheap energy, cheap food, cheap labor, cheap land.[4] The colony was a site of cheap resources, in the sense that they would all be extracted faster than they could regenerate, borrowing "on credit" as it were, from both the land and also from the social organization of the colonized peoples. Operations of extraction helped transform black slaves into living minerals, into commodity-producing commodities.[5] Resource extraction by the old empires, then, was fundamentally an operation through which bodies were disciplined, exploited, and dispossessed in the production of value and goods. Extraction on the cheap continues the depletion of the colony, leaving vast horizons of sweat and blood, toxic landscapes of deforestation, abandoned open-cut mines and their mountainous tailings, soils exhausted by plantation methods of cultivation.

Consider how operations of capital are integral to the production of outsides – those landscapes that beckon further expansion and that nourish the engines of capital itself. Without outsides, capital can't sustain itself, has nowhere to go, and no further bodies to push into its chains of production.[6] Regularities are achieved when the boundaries of capital are constantly undone by the expansion of frontiers, through which cheap nature – in all its encompassing meanings – can be brought into the fold. Operations of capital are thus attuned to experimentation for the ways they rehearse the borders of what can be included within the reach of capital's governmental power, especially for empire.

Yet empire is not a static entity or organizing principle. It is an assemblage of dynamics through which territorial ambiguity is produced in conjunction with legal categories of belonging and exclusion.[7] Its distribution of power, priorities, demands, and violence are embedded in the debris and remains of the post-colony, and its durability stays with the most precarious. Or, to quote Ann Stoler's reading of Frantz Fanon, the depletion economy is a form of power that "slashes a scar across a social fabric that differentially affects us all."[8]

The Entangling Undergig

When we talk about the undergig it is seldom a "we," though it is poverty on all sides. It is those who are affected by the spiralling of poverty in a seemingly circular way – it isn't a clear-cut, linear flow of capital, or dominated versus non-dominated. People rendered "down there," "below" are poor and offshored, and the people in the gig economy are also poor – not necessarily poor in the same way, but caught up in the same spiral of poverty proliferation. Being female, poor, and non-white greatly increases your chances of being employed in undergig work, perhaps earlier than others, but this "sunken place" of digital labor is capacious; white men can find themselves there too. Poverty spirals and tightens the screws around those pushed to the limits of life. Poverty is a condition that legitimizes across class and rotates and feeds back into privilege. The promise of the gig economy is that you can find a "real job," that you can pull yourself up without being entrenched and stuck, while being ever more "screwed."

The undergigged may be continuously employed, but are often in untenable, sometimes invisible, exploitative conditions that underwrite and enable the precarity of gig workers. And because they fall outside the Marxist critique of gig economies, they are harder to imagine as an organized body. Gig workers, on the other hand, can participate in freelancers' unions such as new platform collectives. This does not mean that gig work is not exploitative. Rather, very exploited non-gig workers play a vital role in creating the technologies for gig work to exist.

The undergigged are also the unwilling or uncompensated participants of operations of capital. The ones that make myriad artifacts and infrastructures possible yet derive no benefits but maximum hardships for their precarious labor. They are the migrants, abused children, and the criminalized dead whose images are used to train facial recognition technology so that you can unlock your smartphone with a furtive glance.[9] They are the black San Francisco homeless whose bodies are mined and appropriated to make facial recognition software more inclusive (and surveillance more accurate).[10] The monetization and dispossession of already precarious lives are foundational for the security state. Facial recognition, after all, is a technology designed to know again, to recall to mind and identify the countenance of the Other. And this process makes the Other visible, exposed to practices of control and operations of extraction. The undergigged are a seemingly vast mine in the reproduction of capital and the state, and their management of populations.

The undergig is reproduced through shifts in geopolitical relations, mutations of oppression and within regional national borders. Digital economies preserve this dislocation, but in industries such as call center work, content moderation, and other digital outsourcing. Digital objects travel while leaving the lives of the workers behind,[11] and the undergig workers entailed in producing these new commodities, the commodities that enable gig work, may themselves be rendered invisible in critical explorations of the novelty of the gig economy precisely because their labor appears anachronistic.

The depletion economy works through the production and reproduction of the undergig. Take, for instance, Walmart, often thought of as the "world's biggest firm." We might think of Walmart as producing one particular "underclass," the flex workers employed in its warehouses, in its delivery services, in its supply chains. Beneath this work we might also think of the workers employed in the factories that produce the end-consumer products sold on Walmart's platform, the engineers and designers in Asia employed through flex jobs without health insurance, or of the workers at the palm oil plantations of Indonesia, with labor and land both harvested for the production of cheap goods that in turn "feed" those Walmart employees. All these employees represent participants in the undergig, and as Lily Irani has suggested, rendering this labor invisible itself does a lot of cultural work for platforms that are not simply coterminous with the tech corporations of Silicon Valley.[12]

The undergig is created through the exploitation of asymmetries in power traditionally described using problematic sets of categories consisting of two to three interrelated terms: within the border and outside, the global north and global south; first, second, and third worlds; developed and developing nation-states; formal and informal economies; capitalist and non-capitalist modes of production.

In other words, capital integrates and re-integrates workers into the labor arrangements that reproduce social life as well as produce and facilitate the accumulation of social wealth. But, the labor of the undergig is labor that is undervalued, underappreciated, and most often unseen; it ensures enough to sustain the lives of those it employs in conditions of impoverishment.

It is critical that we move beyond the perspective of nation-states and nationally denominated capitals as basic units of imperial world order.[13] The undergig pushes us to think about processes of differential inclusion across sites, contexts, and actors. It ties in together an assemblage of entities and modes of acting that are not territorially confined by the already porous borders of the nation-state, nor are they neatly separated capitals. We complicate the view that there is such a thing as inequality and exploitation along clear class, gender, and racial lines or along geographical divides, say north or south, or West and non-West, but without sacrificing a critique of the ways in which inequality and exploitation are rendered more intense for bodies that are forced to inhabit the precarious sides of these axes. The undergig is the unbearable weight of contemporary life, a spiral that stretches ever outward in the consumption of lifeworlds ripe for the taking but stubborn enough to threaten its perpetuation.

The undergig contains within it an older, pre-gigification, pre-platformification type of labor: resource extraction, factory work, electronics production in Asia (in turn often reduced to China, which in turn is understood through the trope of Foxconn). There is an inherent assumption in this logic that there are geographical and temporal differences that keep these forms of exploitation separate, while interdependent; Asia (or the global south) has "old," backward forms of labor exploitation, from a previous era of industrialization, which "newer" forms of labor exploitation that occur in the West, the "global north," rely on – an echo of critiques that subaltern studies scholars have made with respect to political and capitalist development in the third world.[14]

While the asymmetrical distribution of these forms of exploitation is undoubtedly true, like our post-colonial studies forebearers claim, our aim here is to "step sideways" and out of tropes of "linear progress" of labor exploitation. Specifically, we are interrogating how the (albeit tainted) endurance of the promise of the "good life"[15] legitimizes the proliferation of exploitation and poverty in a spiralling fashion, back and forth, in and out, abusing (rather than flattening) temporal or spatial claims of difference.

Companies making up the gig and platform economies of digital capitalism distribute depletion and enrichment, sometimes in ways that follow older patterns of empire and sometimes not. They extract surplus from undervalued labor in depleted rural America, or in prisons, or in India, to tag and sift through images for machine learning. Or perhaps they use artificial intelligence, running on server farms close to power and water in cooler zones close to the Arctic Circle in order to figure out an optimal way to design "sustainable" palm oil production in Indonesia. This is the proliferation of the fragmented undergig. The gig hidden by distance, disproportionately performed by precarious women, people of color, far-away farmlands and forests, and animals whose lives are made even more miserable by precarious workers' need for cheap meat, cheap food, cheap sustenance.

Big Tech outsources on-demand code work in India and rolls out facial recognition technologies in Singapore's prisons because these places are already very precarious for coders and inmates. Algorithms don't do work by themselves; they depend on the bodies and expertise of precarious people. Your computer software can schedule a ride, but a human must build the car in Flint. You can click "purchase" for new shoes, but somebody has to dig that coal for free shipping (Amazon ships 1 million items a day). Algorithms can detect child pornography in a Facebook photo, but a human worker in Delhi must witness child abuse in order to tag this content as too traumatic for viewers. The growing undergig endures pain for the ever smaller percentage of *overcommons* to feel pleasure.[16]

And it's not as if the old shit-jobs disappeared. Rather, they are being supplemented by new shit-jobs that are also unreliable. That is why they call data taggers and image classifiers "data janitors" – they clean the shit out of your digital life. If the export is fewer disturbing images on Facebook or cheaper coal, what's being imported is dependent on an undependable kind of work – digital precarity and the undergig.

Techno Toxic

Electronics manufacturing is toxic and disabling. Some of its most crucial tech manufacturing tasks are still done by hand. Undergig labor in electronics production is intimate, often metal to skin. (And yet, paradoxically, it is the human that is rendered toxic to the tech: key manufacturing steps for electronic devices have to be executed in clean rooms, with workers shrouded in protective garb.) Though automation has taken over some of this work, it has not taken over *all* of it; the scale of demand for circuits and the devices that use them keeps pace with the need for human bodies and human capital to fuel the economy of anticipation, growth, and expansion.

Even the "good" digital jobs (those in the tech industry that are non-toxic, well-paid, with the possibility of advancement) are still precarious; like the good life, good jobs now have a phantasmatic quality defined by frequent shuffling and layoffs. And these are the jobs that people struggle to get and keep. The worker's body pays the price; stress, disease, mental health all take their toll. The depletion economy produces massive amounts of disability all along its circuits. It also produces vast amounts of toxic detritus.

More than the electronics themselves, toxicity is the tech economy's biggest export. This is why we turn to how toxicity – the spread of environmental harm and vulnerability in the depletion economy – is the condition for digital production today.

Toxicity operates as a metaphor. Dominant video game cultures propagate "toxic masculinity." Abusive CEOs at start-ups perpetuate tech's "toxic" workplace culture. Properties held by struggling financial institutions might be "toxic assets."[1] These rhetorical uses signal how the idea of bodily invasion and infection weaves in and out of our contemporary technological discourse, even as digital technology purports to be disembodied, smooth, clean.

Yet it is more than just a metaphor. Toxicity burrows deep in flesh as people breathe in its evanescent particles. Brett Walker's deep ethnographic work on heavy-metal poisoning in Japan teaches us that industrial toxins have no boundaries – their traces can be found in deaths from insecticide contamination, poisonings from copper, zinc, and lead mining, or in the congenital deformities that result from methylmercury factory effluents.[2] They all precede and underwrite the age of digital dominance.

What could it mean to think about toxicity as both the production of disabled bodies and as a potent figure for understanding the subject? The purging of toxicity is imagined as a way to re-establish the purity of the subject. Just as toxicity is everywhere, so too are attempts to purge the body of it: juice cleanses, water cleanses, herbal cleanses, colon cleanses, digital cleanses. But as Alexis Shotwell reminds us in *Against Purity*, there is no easy way of immunizing ourselves against our impure pasts and complicit presents.

> Many of us are settlers living on unceded native land, stolen through genocidal colonial practices. We feed domestic animals more food than starving people lack, and spend money on the medical needs of pets while eating factory farmed meat and spraying our lawn with pesticides that produce cancer in domestic animals [...] We cannot look directly at the past because we cannot imagine what it would mean to live responsibly toward it. We yearn for different futures, but we can't imagine how to get there from here. We're hypocrites maybe, but that derogation doesn't encompass the nature of the problem that complexity poses for us. The "we" in each of these cases shifts, and complicity carries differential weight with our social position—people benefiting from globalized inequality are for the most part the "we" in this paragraph. People are not equally responsible or capable, and are not equally called to respond.[3]

There is no cleanse for precarity or for our particular roles in sustaining it.

Toxicity Every Step of the Process

It's no secret that the purchasing of digital devices funds worker and human rights atrocities. Our devices rely on labor and materials that support structures of exploitation and violence. Each step in the production process exposes workers to a different form of toxicity.

It begins with instrumentalizing the periodic table; rare metals are essential materials for ever smaller and more powerful digital devices. The map of rare metals changes the geopolitics of where mining happens as nations scramble to control the extraction of crucial materials. New metal mining industries graft onto formerly colonized landscapes (Latin America, Africa, Australia); they engage the bodies of miners exposed to these metals as countries race each other to control these growing markets. The production of electronics, moreover, requires the mining of high-value raw minerals – gold and the "3Ts": tungsten, tin, and tantalum. Digital devices on the shelves of your closest Amazon warehouse or sitting comfortably in your pocket are possible thanks to the entanglements between metal mining industries and the enduring detritus of imperial refuse.

Next, we might think about how the assembly of devices also endangers workers by exposing them to toxins. Supply chains and the global assembly line converge on Asian (often women's) bodies as they assemble toxic components of high-demand devices. Their reproductive, life-giving capacities pay the price for ubiquitous electronics. One of the richest companies in the world, Apple, has the highest industry mark-ups, made possible by the labor of workers who get sick, lose the capacity to bear children or have borne children with serious disabilities after working in their factories. The depletion economy exports toxicity to import cheaper devices to parts of the world privileged enough to purchase them.

Recent projects of upgrading the manufacturing industries in the coastal regions of China are aimed at taking human inefficiency out of the loop. While China is busy upgrading its factories into high-tech plants, the US strives to "bring back" the slogan "Made in America." US industrial towns are importing toxicity. Under pressure for reelection, politicians are calling for a return to a manufacturing economy. In 2017, the city of Janesville, Wisconsin campaigned hard to attract a Foxconn factory to its small town to replace a shuttered Ford plant that had employed unionized workers. This plant was never built, but had it succeeded, the kinds of jobs that these workers would have had would have exposed them to toxicity. This promise of a return to "Made in America" sits side-by-side with the desolated integrated circuit factories of California, left behind as vast Superfund sites.[4]

After consumers purchase and use electronics, they interact with platforms such as Facebook, which in turn rely on low-cost, vulnerable labor to perform the chronic and grisly task of content moderation. The labor of content moderators involves having to watch and remove toxic content - sexually graphic and violent material (images or references to pedophilia, necrophilia, animal abuse, beheadings, suicides, murders, etc.). Facebook's basic moderation is typically outsourced to countries like the Philippines and India for their familiarity with Anglo-cultural norms as a result of a history of colonization and their willingness to accept extremely low wages (on the order of $1 to $2.50/hr).

Machine learning algorithms designed to detect violent, illegal, inappropriate, and disturbing content online do not simply and automatically remove media containing child pornography from the internet and keep moving. Instead, such media are sent to content moderators who determine whether or not flagged content did in fact contain instances of child pornography. Social networking services make poor and poorly paid workers in the Philippines and in India moderate violent or graphic content online.[5] They often do so to the detriment of their own mental and emotional health, so our timelines and feeds can remain relatively innocuous. This kind of unregulated and non-unionized work is not foreign to workers in the US. More than 5 percent of workers there rely on this kind of crowd-work from tech companies.[6]

As Mitali Thakor has argued, the incorporation of child-abuse detection algorithms by law enforcement agencies has created a new hybrid machine-labor ecosystem that is not limited to the work of traditional law enforcement officers, but more comprehensively includes algorithms and the computers that run them, computer scientists and programmers, and content moderators in addition to law enforcement.[7] Policing institutions justify marshaling resources for these hybrid machine-labor ecosystems of "algorithmic detectives" through a racially exclusionary appeal to child innocence in which innocence is always conferred to potential victims in photographs through a subtle, almost unconscious evaluation of their proximity to the figure of an ideal white child.[8]

It is tempting to think of all this in terms of an unequal relationship between the global north, as exporter of toxicity, and the global south, to which it is exported. This would be an over-simplification, but it cannot be denied that surplused populations across the globe are often the most precarious test subjects that serve the depletion economy. And the cardinal split of the globe doesn't quite cut, as the proliferation of toxicity in depletion zones attests.

To examine precarity in our contemporary moment is to be attuned to the damages, the stubborn remainders (or reminders) of modes of power invested in the differential management of human and nonhuman lifeworlds.

The Fairchild Semiconductor Corporation

Let us not overlook how toxicity is also dumped on people of color and other precarious populations internally, within the territories of the global north as well. Such territories and their populations can become laboratories for testing the results of toxic procedures. They are close enough to study, precarious enough to lack the power to object, and yet held at arm's length, internally, in the belly of the supposedly protected global north.

Consider, for instance, that a 1970 study of the correlation between birth defects and radiation, specifically from uranium mining among Shiprock Navajo workers, found that the "association between adverse pregnancy outcome and exposure to radiation were weak," but that "birth defects increased significantly when either parent worked in the Shiprock electronics assembly plant."[9] Similar correlations were found at other assembly plants in California and elsewhere. Epidemiologists knew what the industry didn't want to know: the suffering of indigenous women and babies was part and parcel of this industry.

The laboratory is where people of color, indigenous people, and poor people are. The reservation in the United States is a space where these three identities live together. It is where experiments have been conducted for more than two centuries now. Shiprock's Fairchild plant and others like it were a space for multiple kinds of experimentation on women of color; there was also a uranium mine nearby, and a power plant, operated by Kerr McGee (Karen Silkwood, a white woman who blew the whistle on this toxic industry, suffered from serious organ contamination after working at Kerr McGee. She died in a mysterious car crash after suing the company).

Thus, a Navajo woman who worked at the Fairchild Semiconductor Corporation's electronics plant had a higher chance of bearing a child with birth defect(s) than if she was exposed to radiation. But her chances of this occurring if she worked manufacturing semiconductors were even higher than that.

It was legal to export toxicity to Navajo women because the plant was built on Native land – because as sovereign nations, the Navajo were not considered part of the US, not subject to the same laws and protections. They were forced to receive this import, in exchange for the export of the tiny components that would power rockets, satellites, calculators, and eventually computers. The State of New Mexico as a whole has high levels of both toxicity and poverty. Some of these poisons were, like uranium, "native," and drove the siting of national labs like Sandia and Los Alamos. New Mexico was a space that technologists, experts, entrepreneurs, the military, and politicians imagined as empty,[10] a place where weapons could be tested and new kinds of labor could be prototyped.

The semiconductor industry in the US first knew about the effects of toxic manufacturing practices on female workers in 1984, when a graduate student moonlighting as a health and safety officer at Digital Equipment Corp told a young new assistant professor, Harris Pastides, that women who worked at these plants experienced extremely high rates of miscarriage. Digital Equipment Corp "agreed to pay for a study" that proved that this was true; three subsequent studies confirmed it, and the results were reported to the Semiconductor Industry Association, which ignored it.[11]

A photo essay commissioned by Bloomberg titled "These Women Are Paying the Price for Our Digital World" shows Korean women who suffer from brain tumors, cancer, and other disabilities as a result of their work at a massive Samsung plant.[12] Their work is the foundation for South Korea's identity as a high-tech nation. The precarity experienced by these Korean women had already happened on Native land in the US, almost ten years earlier.

Reservations have always been economic laboratories, of cigarette consumption and gambling, things we call vices, simply because these products were not regulated there. These areas were made into experimental sites for the digital. From 1965 to 1975, 20 years earlier than the Pastides and other studies, almost 1,000 women worked at the state-of-the-art Fairchild Semiconductor plant in New Mexico, on Native land. These women, along with the thousands of others working in the Fairchild plants in Asia and the US, built the digital industries.

It was the precarity of indigenous women (who all lost their jobs when the plant was taken over by American Indian Movement activists) that created the conditions of precarious workers in the Bay Area. Like their indigenous sisters, women in the Bay Area suffer far more breast cancer than the norm; no one seems to have nailed down the cause.

Silicon Valley exports precarity to places such as Shiprock, New Mexico, the Philippines, Korea, and Malaysia and because the industry is built upon precarity; it shifts locale to where labor is the cheapest and least accountable to regulation. Indigenous women's precarity produces every other kind of precarity in the digital industries. The number of people sleeping in cars, in tents, in RVs – more likely people of color but also the poor – on sites such as the Stanford Campus's El Camino Real, are an eloquent testament to the impossibility of living with dignity in the Bay Area, where real estate is unattainable except by the wealthy, but where jobs as contractors and freelancers are to be had.

The body is a lab for precarious living; precarious bodies are the crash test subjects for the juggernaut of extraction, leaving behind on its trail sites, land, bodies marked by toxicity. Toxicity is never separable from the question of who or what is credited as human.[13]

The Widening Gyre of Precarity

To a person with a hammer, everything looks like a nail. To a person in a laboratory, everything looks like an experiment. Experiments can vary in scale and site. If the colony is a major site of experimentation, so too is the city. Particularly those parts of the city that are already the home to the most precarious. The publicity of the city as a site of experimentation with precaritization offers the opportunity for retrospective unfolding of precarity's *longue durée* through an inquiry into its compounding and spiraling effects.

Let's think about how a previous generation of industrial technologies enabled experimentation on a city-wide scale, the residues of which are still with us. Let's take this example: Michigan is a key center of the automotive industry. In the post-war period, the car was a component of an experimental production process that re-engineered the space of the city. Less-precarious workers hopped in their cars and migrated to the suburbs, leaving the urban core depleted in terms of their tax base. The divide between those who fled and those left behind was heavily racialized.

Ironically enough, the automotive industry not only caused city-wide spatial experiments that intensified precarity for those already at a disadvantage; it was itself then caught up in a global experiment in the redesign of the supply chain for elaborate manufacturing. General Motors (the Apple Corp of the 1950s), once the world's largest and most profitable corporation, tried to maintain market dominance in the post-war economy through implementing technical experiments with task-automating robots, techniques for scientifically managing assembly line production, and offshoring jobs that were relatively automation-resistant.

It's a seldom-told story that today's networked digital media were born, in part, in places such as the GM factories, which pioneered technological and scientific processes. The shop floor became an experimental space for "solving the problem" of labor, turning the screw on labor's sliver of autonomy. Remaining competitive in the second half of the twentieth century meant intensifying industrial output without corresponding growth in demand for manufacturing labor. Producing more industrial commodities with fewer workers became increasingly generalized as unemployment and its threat haunted the lives of those employees whose jobs and families had relocated to the suburbs in the late twentieth century.

Precarity is a social force, vectors of domination that congeal into the form of a spiral. Through labor-saving strategies that allowed for corporate growth in periods of intense competition between manufacturers, precarity has become increasingly generalized; it has insinuated itself in the lives of even the privileged subset of industrial workers who fled once heavily populated industrial cities for the suburbs.

One should keep in mind, however, that the earliest experiments with the devaluation of industrial production were initially concentrated on the racialized and gendered bodies of already precarious workers.

The "problem" of labor was always also the problem of race. Black workers were an integral ingredient in the auto industry's labor experiment. They were hired to be precarious – first excluded from the unions, then excluded from leadership of unions. As workers they were the were first in, first out, and relegated to deskilled, often dangerous jobs. In a sense they were automation's forerunners. It wasn't the gig economy yet, but this lab tested ways to cheapen human labor by racializing it.

But let us never forget the agency of those workers. In the late 1960s, in Michigan, revolutionary black workers took on their own white union leaderships, their bosses, and their local governments. Their movement was defeated before it could spread very far, and black militancy only accelerated white flight and the depletion of urban cores. But for a while it was black workers who led a counter-movement against the experimental laboratory of racialized labor and precarity.[1]

The city as laboratory and the techniques of experimentation that generate precarity have pre-histories in the pre-digital age. But reading the city as a palimpsestic laboratory notebook, one striated by half-marked inscriptions of precarity's historically fluctuating intensities, allows us to see how sliding down the spiral of precarity accelerates as the screw tightens and the gyre widens. With each new experiment, precarity more intensely saturates the lives of the already precarious while simultaneously drawing more lives into the spiral's on-ramps.

Ironically enough, the depleted urban core, into which black communities were corralled, is now the site of a more recent modality of precaritization; once-depleted former-industrial cities have become sites of gentrification as more information-intensive industries and creative industries have moved into urban spaces and drawn knowledge workers to them. Initially celebrated as an attempt to revitalize once-depleted cities, gentrification's sinister logic has led to the intensification of precarity in the lives of those who were just scraping by. And, that was before the cost of rent and coffee started skyrocketing.

These digital industries treat the city as an expanded laboratory for engineering yet more spirals of precarity. The digital makes everywhere a possible site for experimentation. It is predicated on beta versions that are released and tested so users stumble on errors and glitches to be "fixed" even while further failures are afoot. As "users" we are not only test subjects in a lab; we are unpaid laboratory assistants, working to make experimental network software "better." To be the *user* is also, always, to be the *used*.[2]

The city becomes programmable, and its assistants are reworked as code and feedback into its blank slate. The automotive industry used to experiment on the drivers who bought its products, but consumer rights advocacy limited the degree to which live humans could function as crash test dummies. With today's tech industry, though, the unregulated social-technical experiment is the norm once again and has so far escaped regulation.

We get used to working with experimental tech that fails. Some failures lead to scrambled images on computer screens, others lead to stocks erroneously sold at an accelerated pace by high-frequency algorithms. These errors, so integral to the logics of digital technology, disproportionately impact already vulnerable populations or produce new experiences of precarity.

When the tech laboratory is deployed, precarity begets precarity. A most excellent case to illustrate this is the Flint water crisis, where the city can no longer deliver clean water to its majority-black residents. Flint is a city of the Fordist industrial core. Or it was. The water crisis must be understood as a product of a protracted history of crises of deindustrialization described above, a white-flight and automation-induced spiral of precarity that left the town depleted of tax revenue for investing in public infrastructures and their maintenance and forced the city to incur debts that were impossible to pay.

The Flint water crisis was precipitated by the State of Michigan's decision to place the city under emergency management in 2011. After reviewing the status of Flint's finances in 2010, which revealed the city was operating on a $14.6-million-dollar deficit, an estimated 45-percent increase in net debt since the previous year, Michigan governor Rick Snyder appointed a non-elected emergency manager in December 2012 to handle the city's finances and balance its budget.

The crisis began garnering attention in local news just a few months after the city switched its water source to the Flint River in April 2014, a decision that was supposed to offer only a cheap and temporary source of water for the city. Sourcing water from an industrially polluted river flowing through town, however, was not enough to directly cause the crisis. The decision that made the difference in the case of the Flint water crisis was ultimately the result of the emergency manager's choice to cut further costs by not adding standard anti-corrosives to the river water at its treatment facilities. Without these added chemicals, the highly acidic water surged through the city's decades-old lead pipes, corroding the rust and protective lining that prevented lead from leaching into the water that gives life to the city's residents.

Very soon after the city began sourcing water from the Flint River, residents started complaining about discolored and foul-smelling water, rashes on their skin after showering, and eventually lead poisoning, and the presence of *E. coli* and total coliform bacteria as well as disinfection byproducts in the water supply. When "business-minded" governments like Snyder's steer municipalities towards financial solvency through techniques of austerity, one cannot be surprised that these parasitic state and local governments and their financiers are directly responsible for the health problems currently suffered by Flint's residents.

In 2017, the city of Flint hired engineering consultant AECOM for $5 million to accelerate the implementation of a machine learning algorithm that could predict which of the city's water pipes might still need to be replaced. The algorithm was only accurate 70 percent of the time but initially saved the city resources that it would have spent in investigating every pipe connecting individual homes to municipal infrastructures. The lives of Flint's inhabitants could only be protected seven out of ten times. An algorithm designed to save lives was also pushing others deeper into toxicity and the precarity it accelerates.

The spirals of precarity turn and turn again. As more pipes were replaced, the algorithm began detecting fewer and fewer compromised pipes. So the city had to abandon the algorithm and go back to searching through the entire haystack of pipes. This was a more expensive solution that caused the city to fall further into debt. Precarity that begets precarity is mediated by growing mistrust.

The digital laboratory fuels deindustrialization and produces concentrated zones of depletion, where poverty leaves residents vulnerable to exposure to toxins, where heavy metals circulate through residents' blood streams, making their little mineral deposits in certain vital organs, eroding cognitive functioning. In this case, computers can't inspect pipes as carefully as humans can, but computers are much cheaper. They create more precarity for those humans who would have performed by hand the crucial job of caring for infrastructure. Exposure to lead brought about cascading health problems the effects of which degrade well-being over the course of the human life cycle, which in turn exacerbates poverty, debt, vulnerability, and the conditions of precarity *ad infinitum*: a widening gyre.[3]

Like the turn of a screw, the spirals of precarity tighten the ever-enclosing process of exposure. They suffocate and poison, dispossess and displace historically vulnerable populations. Once white-flight leaves the city with a dwindling tax base, and once resources are extracted from predominantly black and Latina/o/x cities like Flint, city governments and entrepreneurial actors begin an urban reorganization project to attract the creative class. Artists, musicians, and so-called innovation hubs are just the opening salvo in white-return (gentrification) to urban centers, a return predicated on the expulsion of black and brown lives. The underside to the enrichment of a zone is the depletion of another.

Some wounds don't completely heal. They scab and scar. Many are invisible to all but those who carry them. There are certain things surveillance does not want to know, or when it knows, it keeps to itself. Flint community members suffer trauma, suggesting a poisoning that could be as spiritual as it is physical, and an inability to trust its infrastructure despite several studies confirming that the levels of lead and hazardous material in the water are now acceptable for residential use. The community has very little to rely upon except its own covens of care, and sometimes those are just not enough. As we might say, after Samuel Beckett: we can't go on; we will go on.[4]

Automating Abandonment

Efficiency is a specter often beckoned in appeals to automation. The computerization of the welfare state, which began in the 1970s, was celebrated as an attempt to make the system more efficient. A lean, digitally-mediated bureaucracy offers the seductive promise that the state will save public funds, serve welfare recipients better by cutting out subjective decision-making and corruption, and simplify processes as diverse as submitting applications and filing claims.

The political demand from which the welfare state grew was guided by *the principle of incrementally overcoming unnecessary suffering*. While such a principle might once have been the guiding light of a certain kind of reformist, gradualist labor movement, the precarity of life is unavoidable. As we age, the odds against survival grow. There will be suffering, there will be precarity – to imagine otherwise is rather too romantic and utopian.

In much of the enriched world and even beyond it, reformist labor movements struggled to embed this kind of ethical socialist principle in forms of administered and institutional care that collectively came to be called the welfare state. Such investments in public care led to the construction of national health systems, but also public assistance for aged care, disability care, and so on. It wasn't restricted to the elderly either. In some countries, it extended across all life stages, from birth through education, work, retirement, and death.

The weakness of the American labor movement, rife with racism, meant that this program was not implemented all that thoroughly in that most enriched of rich worlds. Still, mass-scale social engineering projects such as the New Deal and the Great Society sought to alleviate poverty by supplementing the wage with entitlements, and even providing the wage itself in the case of unemployment, retirement, or disability.

That said, Precarity Lab is not particularly nostalgic for the post-war social welfare state. Like most institutions, those of the social welfare state embodied mixed and even antithetical agendas. Social welfare functioned mainly to reproduce labor so that labor might reproduce capital. It was riddled with racial exclusions. It insisted on normative models of gender and sexuality. It enshrined the patriarchal family model as a norm. And – no surprise – welfare states were also laboratories. They were sites of experiment on the physical and mental attributes of bodies. Welfare disciplined potential recipients by making eligibility conditional on adherence to respectability, a heteronormative family structure, and work requirements. Administrators and experts served in gatekeeping functions, obliging precarious people in particular to ventriloquize the comportment and language that would open the gate to housing assistance, a scholarship, or access to medical gender transition.

Race, gender, ability, and sexuality have been used to distinguish between the "eligible" and "ineligible." The safety net always had holes big enough for certain categories of people, already living precarious lives, to fall through.

Two generations of austerity governance eroded even these compromised forms of the social welfare state in those parts of the enriched world that might somehow have afforded them. No longer is the incremental elimination of unnecessary suffering the guiding light for all to see. But there may be more going on here than just the neoliberal shift to market-based solutions and austerity rationing.

These days, one might take an even less "charitable" view of the purpose and function of the social welfare state, particularly given recent experiments with technology designed to update and automate the management of public assistance. *One could even ask if what is left of the social welfare state no longer even has the mission of eliminating unnecessary suffering, but rather of using digital surveillance techniques to exacerbate suffering as a means of control and rent extraction.*

Automating the Sluggish Inefficiencies of Bureaucracy

Algorithmic decision-making replaces the human, and sometimes more humane, discretion of state bureaucrats such as caseworkers and claims processors.[1] Humans are bad at calculating data compared to machines, but they can be negotiated with – they can have compassion. This is not to say that they always do – their decisions may not be always less racist or sexist than algorithms – but they can.

Yet these fantasies of the technological "quick fix" to the expensive sluggishness of government decision-making obscures a fundamental question: efficiency for whom, and at whose expense? The automation of public benefit administration acts as a covert austerity accelerant that hollows out social programs, while allowing a shell of the various programs comprising the welfare state to persist in name alone.

A US Department of Agriculture fraud detection algorithm recently determined that a New York City grocer was processing food assistance payments after already providing items from his store to customers on credit.[2] The grocer would allow community members to get groceries on credit when they had already exhausted their benefits, and charge them for what they needed to scrape by until their benefits were replenished for the month. The grocer was not lauded for providing community members with foods necessary to survive, and instead was barred from participating in the SNAP food assistance program altogether. Not only did this decimate the grocer's income, but also broadly affected the lives of low-income members of his community. Here, the undergig that enables benefits to be distributed without record and surveillance is punished for providing care, a necessary supplement when social welfare is utterly broken.

By automating bureaucracy, "benefits" can act as techniques of extraction in neoliberal sheep's clothing, covering over the sly dismantling of the welfare state under the guise of efficient technocratic management. All too conveniently the digital technologies designed to render bureaucratic labor obsolete reproduce existing structural inequalities at best, if they do not kick people off of public assistance programs altogether.

The Subject of Public Assistance from Eligible Recipient to Beneficiary

The disciplining function of determining who is eligible for benefits also continues to persist under conditions of automated efficiency. Our allure with disrupting bureaucracy with tech-driven efficiency distracts us from the impact of a system that decides whether or not real people are "eligible" or "ineligible" for receiving public support on top of layers of abstraction that are ultimately reducible to a human-free logic to render binary profiles "ineligible."

Machine learning algorithms might give the binary a valence of probability, but ultimately a threshold of resemblance determines whether or not one is included within the set of the eligible. Because artificial intelligence is ultimately just computationally juiced-up statistical analysis, one is at first never entirely eligible or ineligible. Instead, data about individual behavior is determined to correspond more or less to predetermined statistical models of eligibility and ineligibility. If one's profile corresponds to the eligible model beyond a certain degree of probability exceeding a minimum threshold, say, 95 percent, they will be counted as eligible.

Already precarious "beneficiaries" are often kicked out of welfare programs and disallowed from negotiating with the state offices that claim to serve them. Transferring authority to make decisions about public assistance eligibility from people to machines makes negotiating with those accountable for making these determinations seem impossible. The person abandoned by the automated welfare state confronts a new object, the clunky state interface, and is consequently no longer a subject precisely because they have been barred from predication, from conversation and negotiation – most often, not for the first time.

For example, if a Medicaid patient makes it through the eligibility elimination round of algorithmic determination, she faces the patient portal. Patient portals and online forms are maddeningly and terrifyingly exacting; they require specific types of browsers or apps that don't work well on specific types of phones; tiny keyboards don't lend themselves to typing strings of numbers; bodies already burdened with urgency – pain, debility, collections – aren't easily capable of executing this difficult dance, which must be done perfectly every time. The tiny link at the bottom of the page that invites the sick and frustrated patient to get help solving problems using the site leads to an email address or phone number.

Appeals to the grace of overburdened state employees often turn out to be another inconvenient email in a bloated inbox that is slowly addressed; calling these phone numbers places one in a seemingly endless queue, waiting anxiously, feeling as if the welfare apparatus will never respond to calls for help.

As such, private health insurance is not protection. Patients may abruptly be denied coverage of a specific medication or doctor's visit on the grounds their paperwork was not filled out perfectly, or a doctor that was listed as "in-network" is actually "out-of-network." And the denial of service – a distributed denial-of-service attack, if you will – affects the would-be service providers and the would-be served; in some cases, human doctors and nurses who have trained for years to protect human life and address suffering fight to get to patients just as hard as we fight to get to them for care.

Accomplished by using opaque algorithms to flag people as "non-compliant," the shrinking of the pool of eligibility by welfare offices moves us towards welfare programs without welfare recipients. The welfare state hasn't gone away, exactly; it has merely been hollowed out by the implementation of streamlined bureaucratic obstructions.

Loan Forgiveness

A reminder: precarity is associated with the act of begging for what is necessary for one's survival. The phrase "loan forgiveness" has precarity built into it, for the state is framed as the magnanimous absolver of the borrower's debt. Yet in the case of public-service loan forgiveness, it is the indebted borrower who promises their labor to the state in the hope of eventually being free from debt.

Consider, for example, the promise of education proffered by student loans. Created in 2007, the Public Service Loan Forgiveness program in the United States enabled full-time public-sector and non-profit employees to have their student loans forgiven after ten years of payments. To save money, the US Department of Education contracted FedLoan to manage the PSLF program and repayment plans. All those enrolled in PSLF repayment plans were automatically switched to the private student debt management corporation, which is now being sued for misleading and exploiting borrowers; 99 percent of borrowers who applied for loan forgiveness under the PSLF program were rejected, often because they were tracked into repayment plans that rendered them ineligible for forgiveness, or they were rejected on technical grounds.[3]

The selling-out of educational loan forgiveness, intended to reward the use of education for public service, to a for-profit company, illustrates the irony of the turn to technocratic governance. Automation of social welfare program benefit-analysis has been sold to the public sector as an easier, time-saving alternative to bloated bureaucracies. In reality, automation multiplied obstacles and shrank "eligibility" on technical grounds.[4] In addition, automation demanded a kind of relentless and time-consuming labor of countering the effects of being spit out as ineligible, labor for which many public-service-sector workers have no time.

Lured by the false promise of uplift through education, the worker laboring to pay off an often-predatory loan required as a condition of getting the very credentials required for employment by the state is led to take debt by unclear information. Workers are being screwed over by the far-from-benevolent state, both on the axis of the wage and the axis of debt. The spirals of precarity tighten with the turn of the screw.

Medicaid

Medicaid, part of that "third-rail" of American politics that is deemed untouchable and yet covered in cost-cutters' fingerprints, provides a similar example. As part of the Affordable Care Act (ACA) in 2010, the federal government gave states the option of expanding Medicaid to legal permanent residents under 65 years old, whose incomes remained less than 133 percent of the federal poverty level (FPL). However, the ACA also allowed states to experiment with their own processes of providing public health care to their residents. In some cases, this might involve imposing cost-sharing requirements on those who already have trouble making ends meet, and in other states requiring Medicaid recipients to prove that they are spending several hours per month either working or acquiring training that would allow them to more easily find a job.

The State of Michigan, for example, refers to those who are on Medicaid as "beneficiaries" rather than "recipients" in all of its official communications. We would not refer to lab subjects or experimental animals as "beneficiaries" because they are providing data to enrich institutions. Medicaid has many lab-like characteristics, including automation and experimentation on vulnerable subjects.

States with Medicaid waivers experiment with methods to create a welfare state without welfare recipients. This is not governance by algorithm, but governance under cover of algorithm. As described earlier, Medicaid is not a benefit, but a process for refusing benefits. While Medicaid work requirements appear to target all Medicaid recipients, they primarily single out and target the 6 percent of Medicaid recipients who are currently out of institutionally recognized forms of work; 62 percent of recipients already work full- or part-time, and the 32 percent that are unable to work are in this category due to "qualified" disabilities or obligations related to school or caregiving. Unemployed people who live in rural counties or former-industrial cities that simply lack the jobs necessary to sustain their populations are likely to lose insurance. Moreover, those who do find employment, and work 40 hours a week at the federal minimum wage, will already make 125 percent of the FPL. A small raise or scheduled overtime thus puts recipients of public benefits at risk of losing their insurance.

Medicaid work requirements, along with other related technical "solutions" to the supposed crisis of the welfare state, co-constitute the change in the conceptualization of "welfare" – from a social right to an earned, and in many cases unearnable, reward. Technocratic governance has been sold to us as an easier, time-saving alternative to bloated bureaucracies. Instead, it has led to the multiplication of obstacles and the shrinking of "eligibility" on technical grounds. Anyone navigating the health care system, the welfare system, or student loan services today is intimately familiar with the endless quest to remain eligible for these programs by meeting requirements, acquiring and submitting accurate personal information, or correcting errors in their profiles.

The automation of the welfare state seeks to remove government workers and beneficiaries from the social safety net system. Automation embeds in the system a recourse to passive-voice deniability. "Your health care visit was not approved." "Your public loan was not forgiven." Automated decisions and automated statements pile on each other as if emanating from a faceless, ethereal substance. Within such technopolitics, higher education and health care are not rights: they are unexecutable code.

Who benefits when caseworkers are fired and no one is getting health care and/or is unable to pay off loans? Insurance companies, people in high tax brackets, private loan companies, and above all the massive digital services companies such as Accenture and SAP that program and maintain these systems of digital deniability.

Keeping People Alive in Order to Extract from Them

Exclusion-by-automation is only one aspect of the widening holes in the safety net. An ever more parasitic state and a predatory credit system is obliged to keep people alive in order to retain them as host bodies for extraction. While acknowledging the violence of exclusion from systems of basic life-sustenance, we also note the way that, at least in a US context, the privatization or private "partnership"-ization of state services renders "beneficiaries" as raw materials for profit.

The "innovation" of today's mode of governance is not just to contract with private corporations to administer state programs, but actually to create pathways for the contracted theft of public resources. Private "administration" or "partnership" models of governance emerge out of a belief that state actors are ill-equipped to manage complex human systems such as health care, telecommunications, or mining – a belief, it should be noted, to which we are sympathetic, but from a different ideological position! – and that taxation of the owning class is immoral (a belief for which we have no sympathy).

A brief list of such "public" goods that have been transformed into captive markets for corporations *through state consent* include student loans, the transformation of public housing in Section

8 vouchers, state funding for charter schools, the expansion of the provision of "managed" mental and physical health services through private health insurance companies, and the fusion of shareholder-driven extraction and transmission investment and customer-facing utilities provision. As such, state services such as Medicaid or notionally state-regulated industries such as electrical utilities become clients/captive buyers for corporate actors.

In the absence of pressure from powerful social movements from below, the social welfare state has become a parasitic public built on the backs of people of color and women. But in the age of managed care, the privatization of education, private student loan collection, and private welfare administration, the public no longer exists. Through contracted theft, the managed state turns the "social safety net" itself into a site of extraction. It is a mechanism for transferring wealth to the private sector and funding expenditure in a fiscal climate where government bodies can no longer generate sufficient revenue through taxation.

States fail to generate revenue through taxation; increasingly, they actually transfer revenue smoothly into the private sector. Tax refunds are a lifeline for the 78 percent of Americans who live from paycheck to paycheck. Yet, in 2019, the federal government seized $3.3 billion from student borrowers' federal tax refund for unpaid student debt, up from $2.3 billion in 2016 – a $1-billion increase in a short period of time. This reveals the degree to which parasitic governance is accelerating. While "permanently disabled" borrowers are eligible for student debt forgiveness, private debt collectors such as Navient create obstructions that make it difficult for indebted disabled people to discharge their student debt.

Corporate actors say they are acting in partnership with states. They say they are losing potential earnings by agreeing to offer services at fixed costs and foregoing the potentially lucrative vicissitudes of the open market. What actually happens is that these managed/administration arrangements create captive buyers – in the case of prisons, literally captive – which insulates private companies from risk and guarantees them not so much profits as rents.

Regardless of what they say to the public or to the state, these same companies tell shareholders that they can expect a stable return on investment, and insulation against loss of market share or collapse. This is a mode of governance whose mission is the welfare of the corporation. It is the corporation and its institutional investors who are shielded from the particular form of precarity that is the "free" market. Even if services are offered *at cost* – which, for the record, they almost never are, as private actors are generally permitted to establish "fair" margins of return in exchange for their "labor" administering state services – they still bolster the corporation's earnings. States participate in this transfer of wealth by charging consumers, or by funneling consumers into particular fixed marketplaces, while the corporation earns the benefits.

The human being who seeks medical care, who wants to call a family member while incarcerated, who wants natural gas to heat their home, who takes out a student loan to pay for school – that person is the raw material from which corporations extract secured rents, all while dressed in the clothing of the state providing a safety net. The automation of the welfare state uses the ruse of "efficiency" to dismantle the so-called social safety net, producing precarity while claiming to provide security.

Fantasies of Ability

Picture Tony Stark, American boy-genius tech entrepreneur, kidnapped by terrorists, wounded in the chest – but who invents for himself an invincible power-suit of armor. He becomes, in what is now a delightfully anachronistic act of naming, Iron Man. His disability is that he has no heart, but no matter, tech and capital come to the rescue, and he goes on to save the world from evil, and all that.

This able-bodied fantasy of invincibility disavows death and sickness as the underlying conditions of existence. Able-bodied subjects can react with fear and disgust in response to disabled people who remind them of the inevitability of disablement and death; contact with disability is a momentary rupture of the psychic barrier that shields the able-bodied subject from knowledge of its corporeal vulnerability. Able-bodied people manage their own precarity by death denialism.

In the enriched zones, such death denialism is made possible in part because sickness and death are not equally distributed. The body is an index of social inequality. Global supply chains and domestic sites of precaritization rely on the immiseration of disposable bodies: premature death from exposure to toxins, overwork, injury, and strained mental health. Workers who perform labor in the most dangerous parts of the supply chain are vulnerable to sudden and spectacular forms of injury.

Slow death, to borrow Lauren Berlant's term, accrues to subjects who absorb a constant stream of blows that, while they may at first appear imperceptible, accumulate over time.[1] In a similar way, Rob Nixon has used "slow violence" to describe the gradual and often invisible toll of environmental crisis on the poor.[2] For a farm worker, this can be the delayed onset of cancer caused by prolonged exposure to Monsanto's Roundup Ready herbicide, or for a black woman who exists under the conditions of atmospheric anti-blackness, it might be manifest in a depressive episode caused by chronic stress and elevated cortisol levels.[3]

From Each Beyond their Abilities; To Each According to their Precarity

Bodies are hierarchically ranked by their laboring capacities. The disdain towards disabled people reveals the extent to which the category of the human is defined by the ability to perform waged labor. Disabled people are scorned as work-shirkers, as if one's capacity to work were the only thing that makes one worthy. As historian Sarah Rose has shown, the rise of industrial capital in the United States was coextensive with the rise of "vocational rehabilitation," reframing disabled subjects as "idle" leeches and therefore necessitating their mandatory re-entry into work.[4] Yet it is that industrial capital itself that legitimizes disposability, operations within which accumulation begets dispossession, extraction begets depletion, and bodies are treated as instruments of productivity. Precarity leads to premature death.

We must not accept the uneven distribution of bodily and psychic harm against precarious people. How do we acknowledge the violence that extraction does to our bodies? How do we avow and embrace what Saidiya Hartman calls waywardness, the refusal to function, and the beauty of brokenness without romanticizing the system that does the breaking?[5] From "wages for bed rest" to crip dance parties, we defend our right to aberration, to our belief in the vibrancy of what veers but is never vanquished.

Able-Bodiedness is More Than a Condition

Able-bodiedness is a critical subject position against which disability politics needs to organize. As digital capitalism accelerates, however, we are called to consider "able-bodiedness" otherwise. How do we celebrate disabled bodies, while fighting the uneven distribution of bodily and psychic harm? This question follows from the work of scholars working in what Jina B. Kim has coined "crip of color critique." Troping on Rod Ferguson's call for a "queer of color critique," a corrective to queer theory that centers material inequities that produce sexual and gendered abjection at the same time as it celebrates non-hegemonic behaviors, Kim calls for a disability politics that can grapple with both the experiences of disabled subjects *and* combat structural systems of disablement. Today, disability politics is a (still primarily white) movement of identitarian self-actualization, but it is also a radical reordering of what types of bodies signify as human. The potential and the pitfalls lie hand-in-hand; making death and mutilation seem humane is an ethical risk, as is failing to make human the subjects who suffer at the hands of war, misery, torture.

Indeed, we don't need to look to Silicon Valley for evidence of racist disablement produced by capital; we can look to silicon itself. Muriel Rukeyser's poetry in *The Book of the Dead* exposes the "acceptable" risk borne by workers in extractive industries, the slow deaths by silicosis – tiny particles of silica in the lungs that scar the lungs from the inside out.[6] The Hawk's Nest Mining Disaster, one of the worst and least-known industrial catastrophes, occurred slowly, between 1930 and 1935, but led to rapid death for miners who contracted silicosis; most survived less than a year. A commemorative plaque in West Virginia reads: "Silica rock dust caused 109 admitted deaths in mostly black, migrant underground work force of 3,000. Congressional hearing placed the toll at 476 for 1930–35. Tragedy brought recognition of acute silicosis as occupational lung disease and compensation legislation to protect workers."[7]

And yet, while equally apt to cause bodily harm, today's extractive disablements take place in a different economic and technological world. Consider that the emergence of the twentieth-century welfare state in the United States deemed certain types of presumably infirm bodies worthy of material support in the form of Social Security Insurance, Aid for Families with Dependent Children, and state aid for people with various forms of blindness.[8] While these forms of state benefits reified gendered logics, building on a long tradition of aid for "war widows," and the nation's intellectual hubs simultaneously promoted eugenic and racist logics that sought to eliminate "feeble-mindedness" and other disabilities altogether, the idea that some persons should be supported without the expectation of continuous work does signify a desire to, at least on the margins, provide care to some subjects. In other words, the state desired to prevent the elderly and some with disabilities from simply dying from starvation and exposure. Now, right-wing and center-left states are shrinking these types of programs at the same moment as digital technology promises to make more jobs "flexible" and "adaptive." This is not disability liberation.

To be clear, the rise of prosthetic technology assists in individual mobility and entry into the social for many people with disabilities – we are not against prosthesis. Nor are we against medical intervention *per se*. Indeed, many of us ourselves are beneficiaries of medical technologies that allow us to live in the world, including access to gender transition technologies that circulate via the same global supply chains that we critique here.[9] (Another mark against a politics of purity: in our bodies themselves live exogenously produced chemical substances, hormones, that some might consider toxins.)

What we do note, however, is the normalization of the white two-income household, the emergence of work programs for people with disabilities, "welfare to work" frameworks from the 1990s, state experiments with adding (as of this writing, questionably legal) work requirements to Medicaid, which taken in aggregate reveal a turning away from the idea that some bodies are not necessarily mandated as laboring bodies.[10] The risk of celebrating digital technology as a sort of universal prosthesis, a technology that allows people with disabilities (especially mobility and communication disabilities) to participate in forms of adaptive labor, is that it carries with it the expectation that all people should thus be willing and able to work to survive.

If adaptive technology is co-opted as means to expand the labor market, just as twentieth-century feminist claims of labor discrimination and patriarchal domesticity have not resulted in a stable two-income family but rather in the increased outsourcing of childcare labor and the expectation that only dual-wage units can expect to survive, this is not disability liberation or accessibility of any kind.[11] Instead, it is simply a reification of an able-bodied fantasy: that all of our bodies have indefinite labor capacity, even if that capability might appear in different forms.

As state support for sick, aged, disabled, pregnant, care-working, etc. bodies dwindles, its expectation that bodies formerly excluded from the category of "worker" become workers increases. We resist the digital serving as the fantasy object that enables such an expectation. If technological prosthesis provides access to desired futures for people across bodily and mental difference, let it be a future of accessible play, not merely accessible work.

Precarity is a Threat Against the Right to Live – It Cannot Be Solved by the Right to Work

Precarity not only wears away at the body but also at the soul.[12] Precarity causes trauma and is caused by it. Think of the emotional labor required to make others happy, soothing their moods, anticipating their needs.[13] We find versions of this emotional labor in a range of jobs from client work to teaching to sex work. Precarity pushes workers into forms of labor that levy additional taxes on already overworked psyches.

How is disablement produced through global supply chains? Who is sacrificed? Digital economies contribute to the uncompensated breakdown of the mind and body. The precarious laborers of the digital sit at screens and filter violent or sexually graphic content, mine digital gold, operate tech support, mine literal cobalt/rare metals, assemble chips and screens, enter data, recycle or dispose of digital waste products, clean buildings in Silicon Valley, drive app-based rideshare cars, transcribe walls of text, write code, edit code, write fanfiction, produce videos, tweet, load computer chips on container ships, buy and sell stock. Some of these jobs are physically demanding, from mining to sitting hunched over at a desk. Others are psychically demanding, even traumatizing.

The risks of labor, its likeliness to result in disability, sickness, trauma, stress, death, are organized by race, gender, and other categories that come pre-loaded with precarity. But by placing both disability and able-bodied fantasies of endless embodiment in conversation with digital precarity, we join with disability scholars of color who ask critical questions about the production of disability under capital and the assignment of the status "disabled" to those who do not work according to capital's plan.

This move is in direct rebuttal to the techno-optimist fantasy in which digital technology becomes a universal prosthetic.[14]

The fantasy of the robotic or digitally enhanced body appears throughout strands of popular culture, from *Avatar* to *Detective Pikachu*.[15] The sustaining trope is that people with mobility or sensory impairments will obviously benefit from digital technologies such as virtual reality that allow them to "escape" their "confinement" to wheelchairs and other real-life mobility aids. While prosthetic technologies are not, in themselves, tools of eradication, the persistent fantasy that "tech" is what can "cure" disability is an eliminationist one.[16] Such a fantasy perpetuates the disgust and dismissal of disabled subjects, while also attempting to recruit them as future workers.

If digital world-making includes an ableist fantasy of "transcending" the body at its heart, yet perpetuates disablement through its labor practices, how might subjects intervene? The digital doesn't transcend the body; on the contrary, it requires ever more narrow performances of it. Instagram is now a "social factory" and normative physical beauty is its stock in trade.[17] In this marketplace, even acts of seeming resistance are fraught; reactions to popular TikTok dance performances by disabled people are divided between vocal support and catcalls. Though digital technologies were celebrated as providing new kinds of labor to accommodate disabled bodies, the rise of image-oriented platforms that celebrate and necessitate perfect bodies undermines this possibility.

Or better yet: how do marginalized people bend the digital networks designed for the circulation and organization of wealth and work to their own uses, to scaffold communal care in the face of bodymind breakdown and precarity? How do they tap into the grid and steal a little electricity for their own?

After the Social Safety Net, We Must Create Social Safety *Networks*

Digital care is a life-sustaining practice that uses digital technologies to create covens of care across geographic space. These might be thought of as social safety *networks*, communities that replace inadequate or cruel so-called "safety nets."[18] State-run "safety nets" are traps for people with disabilities, who, if they make too much money or get married, are balanced precariously between "not poor enough" and "too poor to live" and are kicked off benefits.

Some might valorize digital social safety networks as evidence of a techno-libertarian promise – that individuals, left alone with a tool, can self-govern and therefore eliminate the need for redistributive or revolutionary economic policies. We instead view these formations as sites of intervention into capital accumulation and individualism that happen to organize in whatever environments, including digital ones, marginalized people find themselves in. Interdependence, the radical disability concept of mutual life, is the guiding framework behind these practices.

Care is care because it remedies precarity. Networks of digitally organized money transfer such as GoFundMe and Kickstarter, crowdfunding vehicles like Patreon, and subscription sex-work services like OnlyFans are used to raise rent money, medical expenses, and other necessary survival funds within marginalized life-worlds. These are mechanisms through which marginalized people literally transfer funds from one precarious individual or group to another.

However, money transfer systems don't add to the sum total of resources held by these people; instead they shift energy to where it is needed at the time. Like the internet, a network designed to withstand damage and censorship, social safety networks are self-healing, but they can only distribute the same amount that has been put into them. Moreover, most of these transactions are "taxed" by the corporations that own these financial services. They are able to extract a rent from the covens of care that form to heal precarity itself.

Social safety networks are glitches, not integral features, of capitalist digital worlds.

Dispossession by Surveillance

Start with 100 points as a blank, generic human. Deduct from your score if you: do not live in the global north, are a woman or person of color, are trans or disabled or elderly. There's no end to the permutations on how these point deductions can concatenate.

Bodies have to be constantly under surveillance. Have your phone on you at all times, so that it gives off a steady stream of telemetry. Your velocity, your purchases, your moods as registered in your subtweets. It's all data for the mill.

This is a seemingly benign-sounding version of the surveillance that generates rankings. Bodies can also be under video surveillance, with facial recognition. Or under audio surveillance, the algorithm primed to detect some alleged mood change as measured through intonation. At work, every action can be monitored, generating an "inactivity report" if you pause to draw breath. Real-time activity measures can even be fed back to the worker, as if it were a game, with some paltry prize held out for the one who works the hardest.

Maybe you have to wear an electronic shackle. Maybe you have to blow into a breathalyzer to start your car. Maybe you are "known to the police" for supposed gang activity. Maybe.

As you go down in rankings they become measures of abjection. More and more liberties may be taken by others with your body and with those data traces it produced. You are treated as if you do not merit care, as if you do not have a right to live, as if you do not have a right to the autonomy of your own body. As if you do not even have the right to be mourned.[1]

Economic migrants are especially subject to ever-mutating principles of subjection. The border has for a long time functioned as a lab for testing out both high-tech ambient as well as old-fashioned surveillance and control.[2] The condition of the migrant is that of constant scrutiny. Slaves and migrants have always been scrutinized; they have been treated as data assemblages to govern; they have never been invisible. And whenever they were able to hide their presence, actors and technologies worked hard to make slave and migrant bodies into perceptible matter. In this sense, algorithmic power extends and strengthens the prejudices, the alibis for violence, of another era. Algorithmic surveillance became a scandal for those who assumed that they had a right to privacy. Subjects managed by the white-settler states, however, are used to this lack of privacy. In the era of generalized digital surveillance, those who used to feel safe in the dormitory suburb that is the enriched world are getting a taste of seemingly ordinary modes of governance.

The generalization of digital surveillance is bringing to the enriched zones a level of exposure formerly only felt by the most precarious. It is producing an affronted class, led by panicky white boys and Ivy League worthies afraid of losing their sovereignty and privacy. The affronted class shake their fists while ignoring their own complicity in the (re)production of precarity or their distracted glances when confronted with depletion zones.

Take, for example, panic about *surveillance capitalism*.[3] It does indeed seem the case that all of human existence is now processed as the raw material for observation and evaluation. To the affronted class, this appears as an unprecedented form of power, as if it had emerged in the last decade. Indeed, it may have several novel features, though subjecting certain bodies to surveillance is not one of them.[4] Current fears of the "end of privacy"[5] emerge out of the anxiety of the affronted class, eliding the fact that privacy is a luxury that racialized bodies rarely enjoy.

The affronted class becomes nostalgic for a pre-neoliberal form of capitalism that is fancifully imagined as inclusive, responsive to the needs of consumers (workers are rarely mentioned), democratic, competitive, and based on the production of high-quality products. Critics of this or that hyphenated form of capitalism are invested in rescuing capitalism from its crisis of legitimacy by taking as their object of critique neoliberal capitalism, surveillance capitalism, monopoly capitalism, platform capitalism – as if the object of analysis could be captured in a modifier.

There are indeed novel features to how the proliferation of the digital generalizes surveillance and renders more and more bodies potentially precarious. The transformation of the business model first used by tech firms has now been adopted in virtually every other industry. In the product or service-based business model, revenue was generated by fees and the sale of goods. Industries are now committed to a business model based on the commodification of behavioral data and the sale of predictive products to actors from advertisers to the state who have a stake in predicting future human behavior.

Claims that surveillance capitalism's process of dispossession is unprecedented ignore the roots of contemporary surveillance practices in techniques developed within the contexts of slavery, settler-colonialism, and empire. Technologies of observing and tracking workers have developed side-by-side with techniques of control intended to exploit new terrains of accumulation. Colonial and racialized subjects have been objects of surveillance since the advent of chattel slavery and settler-colonialism. Their movements, actions, behaviors, feelings, and bodies were zealously observed, guarded, and documented – not least in case they showed signs of revolt. But there were technical limits to how much this information could be gathered, processed, transmitted, and enacted. The tens of thousands of fugitive slave notices appearing in the American newspapers of the 1800s at least left some margins of time and space through which the fugitive slave could disappear.[6]

In the laboratories of slavery, surveillance was an overhead, a form of friction, resulting from its inherent forms of violence but counting as a cost against the value of the products produced.

What is distinctive about the age of generalized digital surveillance is that surveillance becomes itself a form of control of value production. Surveillance is no longer a byproduct; it is a business model, one that generates its own laboratories of experimentation with the rendering precarious of more and more classes of bodies, reaching up even into the privileged ranks of the affronted class – at least in their panicky imagination.

In settler colonies from South Africa to Palestine, prisons and police are used to both displace and control the indigenous inhabitants. Such conflicts catalyze the creation of new instruments of repression and control. This is a space of experimentation Wang calls the "carceral laboratory." It is a zone where new techniques of control are tested out on society's "others": women, minorities, criminals, queers, the underclass, and colonial subjects. For example, the West Bank and Gaza function as a carceral laboratory for Israel, which then exports technologies of repression to states around the globe.[7] At the same time, laboratories of repression also become test zones for resistance. In Palestine, the laboratory of resistance can be found riddled with incendiary balloons let loose into Israel,[8] with Palestinian youth returning to live in depopulated villages,[9] or in the continuous rebuilding of razed Bedouin homes. The scientists in laboratories are not the only ones conducting experiments – their test subjects *sousveille* as the scientists surveille.

Colonies, from European territorial colonialism to contemporary settlements, have functioned as laboratories where techniques of surveillance are developed, tested, refined, and then converted into domestic policing infrastructure. What was once known as the "McNamara Wall," built and designed by the US military to surveil cross-border movements between North and South Vietnam during the 1960s–1970s, was turned in 1970 into an "electronic fence" that monitored unauthorized border crossers from Mexico now imagined as "intruders." Contemporary calls for "smart walls" hark back to the conditions of possibility set by the "McNamara Wall" and the "electronic fence." Today, predictive software developed by the Department of Defense for counterinsurgency in Iraq and Afghanistan has been converted into the predictive policing software PredPol. Such projects have their roots in the early days of the US's imperial undertakings. In *Policing America's Empire,* Alfred McCoy argues that the modern American surveillance state emerged out of the US's colonial experiment in the Philippines, beginning at the end of the nineteenth century and into the first half of the twentieth century. In addition to repatriating information-based policing techniques, colonial administrators repatriated conceptions of race and methods for dealing with the nation's "internal others": "After years of pacifying an overseas empire where race was the frame for perception and action, colonial veterans came home to turn the same lens on America, seeing its ethnic communities not as fellow citizens but as internal colonies requiring coercive controls."[10]

At the same time that the US was developing domestic urban counterinsurgency tactics to put down black radicals and the anti-war Left, the Cold War prompted the US to export its professionalized police tactics to use against communists abroad. In 1962, President John F. Kennedy established the Office of Public Safety (OPS), an agency that worked closely with the Central Intelligence Agency (CIA) to train police in South Vietnam, Iran, Taiwan, Brazil, Uruguay and Greece. Though this Cold War project was disbanded in 1974, similar projects continued to train agents abroad in security tactics for the purpose of crushing communists and facilitating free trade. Many of Latin America's most notorious despots were trained at the US-operated School of Americas. The military training school was founded in 1946 to secure the Panama Canal Zone, then shifted to the domain of anti-communist counterinsurgency, and whose alibi now, among other things, is the "war on drugs."

Even when used as a tool of governance by the state, surveillance has been key, not just to the maintenance of capitalism and empire, but to their experimental elaboration. The state's need to control colonial and racialized subjects and protect the hegemony of capitalism authorizes the expansion of surveillance and policing.

When considering the difference between technology designed for consumption and technology designed for state surveillance, we see two kinds of branding at work. On the one hand, there is corporate branding as a marker of status. On the other hand, there is a kind of branding where subjects are forced to bear a mark of stigmatization. On the consumer side, digital technologies are coveted when they are tiny, powerful, and expensive.

In 2014, Nexus created a new tracking device meant for detained migrants to use as an alternative to detention by Immigration and Customs Enforcement (ICE). We could call this technology an ankle bracelet, but we prefer the word "shackle," which is how the *New Yorker* describes it. Nexus' GPS-enabled shackle, ironically and cynically called "Libre" (Free), costs $420 per month to rent and is often worn for 25 months, which is the time most immigration cases take to be resolved. If "clients" wear it for the duration of their case, as of 2019, "someone with a seventy five hundred dollar bond would pay Libre nearly thirteen thousand dollars in rental and other fees. (The vast majority of the four hundred thousand people detained by ICE each year are deemed automatically ineligible for bond and remain in custody until they win their cases or get deported.)."[11]

Libre is used to monitor and keep track of the location of detained migrants. The enclosure of the holding cell is extended into the confines of what is meant to be their home.[12] Control and surveillance in this sense is modular as it follows the surveilled in their everyday life. Just as border control was reimagined through its engagements with cybernetics thinking in the 1970s,[13] immigration control is reconfigured by treating migrants as data assemblages to manage and control. The shackle, however, doesn't only monitor and attempt to enclose migrants; it also mutilates their bodies. These devices generate intense heat when charging that then burns the skin.[14] Here is where the electronic shackle does not hide its long-standing connections to the history of the plantation and of slavery. It echoes the execution of branding as a racializing surveillance.[15] The conditions of living under detention go skin-deep for these migrants, underscoring the fact that not all bodies are surveilled the same. Some bodies are rendered markable, shackle-able, while others are not.

The GPS shackle helps transform the bodies and lives of detained migrants – just as it does with parolees – into extractable matter. The monthly rent and fees push those that are shackled into indebtedness while their futures remain uncertain. The *coyote* (smuggler), long understood by border studies scholars as a product of US immigration policy and operations, is the other side of the same coin. The coyote extracts value from the precarious conditions of unauthorized migrants. The coyote profits by helping such migrants navigate the borderlands without being surveilled; Nexus profits from migrants' intense vulnerability as surveilled subjects. Precarity is both a product of capital as well as the grounds for the (re)production of extractable matter.

Immigration has often been both a response to precarity and a cause of precarity. Unauthorized migration is a desperate move by desperately precarious people who have been deprived of their livelihoods by the very industries that create the objects they must mortgage their lives to rent. Those whose lives have been scarred by empire's export of war are trapped in the production of their own displacement and migration.

In the enriched zones, among the panicky liberals of the affronted class, it is mostly consumer-grade surveillance, assessment, and control that generates the opinion pieces. Even white men with jobs and homes who don't think twice about what can happen to them when they walk the streets start to worry about their Fitbit data, or whether Alexa knows their porn habits.

These enriched zone problems call for two kinds of contexts. One is historical. The forms of surveillance creeping into such enriched lives have precedents, and those precedents might be many and varied, but all of them were laboratories that experimented on "the human" to render it precarious in ever new ways.

The other context is to look at depletion zones all around us today, where the kinds of intrusion of surveillance into everyday life extends far beyond questions of keeping one's personal browsing habits private. The precarious, both past and present, are both subject and product of laboratories whose techniques are more and more generalizable in the age of digital surveillance and control.

The Affronted Class

Lately, even the whitest of white boys are finding themselves without security. Automation renders the jobs of once privileged, exempt, insulated classes of technical and creative workers increasingly vulnerable to obsolescence. The magic kingdom of disrupting this and innovating that and pivoting other people's lives out from under them has caught up with them too. Nevertheless, the precarity that these privileged classes are now experiencing is not comparable in degree to those who have lacked such racial, gendered, and economic privilege.

They are not taking it well. Some seem to have the right idea, and think about organizing as workers, or subjecting the final goals of their work to ethical and political scrutiny. Others take their vexation out in counter-productive ways. Some even become *fash,* as it is now the fashion to call (neo)fascists. Some are reacting in even stranger ways, as we shall see.

Think of Evan Peters' Kai Anderson in *American Horror Story: Cult,* a Trump-loving unemployed and precarious white boy who starts a white supremacist cult that attracts women of color, queers, and liberals. His close relationship with his Clinton-supporting sister furthers *AHS: Cult*'s incisive observation: responses to precarity are increasingly strange and illogical, and the liberal Democrat has more in common with a Trump-touting fash than she cares to admit.

Neither a bourgeois nor a proletariat, as Silicon Valley's start-up industry rearranges the very notion of ownership of the means of production, this is a newly *affronted class* that consists of those who thought they were no longer workers – at least not blue-collar, manual workers. They had once shared a sense of being exempt from the relentless logic of exploitation. Now they and their descendants find the good life slipping out of their hands. The affronted class is what has become of the secure workers of the mid-twentieth century, who thought they were experts on how to bunker the good life.

"Tech Bro" Feelings

One archetype of the affronted class has a name, and indeed it is a stereotype. The figure of the "tech bro" is an outgrowth of this new landscape of specifically bourgeoise masculine vulnerability. It is used to describe a subculture of mostly male and mostly white (or white-acting) tech workers and entrepreneurs associated with the masculinist culture of Silicon Valley. White programmers occupy a privileged position within California's economy, a class position that historically emerged in conjunction with white workers' anxieties about indentured Asian laborers in the mid-nineteenth century.[1]

For convenience, let's date the rise of tech bro discourse as emerging post-2007–2008 financial crisis, when the critique of the tech industry's normalization of a technocratic, libertarian, privileged, meritocratic culture of labor exploitation and exclusion along lines of class, gender, and race went mainstream.[2]

The tech bro identity and lifestyle cuts across fields such as computer science, design, and engineering. What is particularly ironic is that many in the affronted class have actively contributed to, if not outright designed, the means of their own demise. Consider how machine learning ultimately displaces the technological authority of lower-rank computer programmers. The programmer feels no longer in control, no longer capable of grasping what he (or sometimes she) creates. One might wonder at this point if even Google understands its own search algorithms.[3]

Old techniques of regaining control no longer work. The familiar strategies of cracking open the black box, fighting the man, pushing against the system have lost their grip. Those once empowered to push back, to make the world anew in their own image, long for what they took to be only theirs: agency, control, authorship, voice, actionability. The playful, daring, untraceable figure of the hacker or the hacker class has been rendered obsolete in spite of the best efforts of the more class-conscious of their kind.[4]

The newly diminished tech worker, imagined as *de facto* white and male, wrestles with the relinquishment of bygone fantasies of modern progress and the security it has long represented for some at the expense of others: the American middle class, the nuclear family, technological promise, economic development. White boys' sudden shock is however *not* a coming to terms with the violent destruction of the modernist project of reproducing the world in their image. Attempts to "solve" the tech industry's new precaritization have produced not a reckoning with its own attachment to a legacy of celebrating progress and technological advancement, but a reimagining of masculinity as either a nostalgic return to traditional values or a neo-colonial project.

The Return to Craft

In the United States, the "maker" is another figure that has grown out of the affront to newly vulnerable white masculinity. The rise of a global "maker's movement," as it has become known, was born in the wake of the financial crisis of 2007–2008. The Western tech industry witnessed a return to craft, a nostalgic longing for what some (mostly men) once had: a sense of agency and control. The spokespeople for makers were prominent figures (mostly men, some but not all white) in the American and European tech industries. Makers, they said, were "returning" to a masculinity and security that had been lost under mass production, deindustrialization of the global north, globalization, outsourcing, and automation. This masculinity was grounded in a deep connection with the machine, a hands-on engagement with technology (and by extension nature, materials, and life itself), and authentic forms of craftsmanship.[5] We see this lost form of masculinity as part of what once empowered men to retain agency and protected their work from feminization during early-twentieth-century industrialization.[6]

This nostalgic return to craft "is not your father's DIY" according to one of its spokemen, Chris Anderson.[7] Yet many invoke a familiar yet distant father figure; what they can no longer hold tight in their hands is what made their fathers manly, masculine, wholesome. This father is of a previous era, an authoritative, benevolent, successful middle-class man, in control of the inner workings of his car and his household, tamed, managed, and loved through principles of rationality, ingenuity, and tinkering. The newness of this DIY is in fact revolutionary for its proponents,[8] because the maker's tools and instruments, from 3D printing to open-source hardware platforms, seemingly give the individuals control, not just over craft but over the means of production.[9]

The nostalgic longing of men like Anderson for a time that once was, to return a sense of agency and control to white boys, fails to grapple with their complicity in colonial and neocolonial projects of exploitation. This nostalgia is fundamentally about recuperating masculine privilege and protection from the precarity implied in the 2007–2008 financial crisis' attack on the promise of technological progress.

The techno elite have responded to anger about precarity generated by the digital economy in the style of their corporate-colonizer forebears. Hands over ears, they refuse to hear the complaints from the precarious; they sweet-talk their way out of accountability, and fantasize about how to escape with their money after their empires fall.

Neo-Colonial Tech Tours and Apocalyptic Escape Plans

For his testimony to the US Senate's Commerce and Judiciary Committees in 2018, Facebook's Mark Zuckerberg wore a suit and tie – as if wearing grown-up men's clothes would prove to the world that his company could be trusted to do some adulting for a change. He talked about Facebook's failure to prevent political consulting firm Cambridge Analytica and foreign powers from influencing US voters. He shared his vision of Facebook as "an idealistic and optimistic company," a company that is primarily "focused on all the good that connecting people can do"[10] – as if this kind of happy talk would make scrutiny of Facebook's vast engines of surveillance go away.

The assumption in Zuckerberg's pitch is that connecting the world's population to one another via the internet, or rather via Facebook, will, hey presto, solve the problems of intensifying precarity. This techno-utopian vision of connectivity obscures the extent to which these connections themselves secure the proprietary platform that enables it and from which Facebook extracts both data and rent. Zuckerberg's attempts to subsume the global population into its branded theme park as if for our own good – for example, Zuck's push for India to adopt Free Basics by Facebook – elides a desire to route ever more user data through his company's servers, capturing information, and analyzing it to improve the platform and increase the value of Facebook stocks. Facebook is exemplary of an emerging model of class rule based on asymmetries of information. You get access to your own friends, or to distraction, or to vital information, or to porn, but only in morsels. Information in the aggregate belongs always and only within the black box of the corporation.

This kind of corporate power through asymmetries of information has to keep growing and growing. Back in 2016, Zuckerberg traveled to Nigeria and Kenya, where he met up with the key actors of what the Facebook CEO called "Africa's emerging IT ecosystem." More happy talk. Zuckerberg played into stories that have proliferated in Western news media outlets and that portray regions at the so-called former periphery now as the rising center of contemporary innovation, hopeful in a moment of doubt over the promises of the Western tech industry, modern progress, and the tech industry's complicity.

From Africa to China, zones of the so-called former "tech periphery" are celebrated for their efforts to follow the footsteps of Silicon Valley. Kenya is dubbed the "Silicon Savannah" for its advances in digital finance, tech incubators, and local IT innovations such as BRCK and the Ushahidi crowdsourcing platform among others. Shenzhen, in southeastern China, is the "Silicon Valley of Hardware," celebrating the region once labeled as backwards and fake, having escaped the West's perils of intellectual property regimes and modernization. There is probably some desert somewhere being celebrated as the "Silicon Silica."

In these happy stories, necessity and lack of resources in the so-called developing world are celebrated as key to the transformative power of innovation. Erik Hersman, an entrepreneur who grew up in Kenya and Sudan, and who refers to himself as "the white African," describes this sentiment in a TED talk as: "If it works in Africa it works everywhere."[11] Regions including Kenya and southern China are rendered as a hopeful toolkit where the promises of Western tech production can be recuperated, despite their backwardness and underdevelopment. Neo-colonial aspirations couple with neo-orientalist tropes of othering.

However, the ways in which technologists are responding to this new-found sense of insecurity are not always so practical. Silicon Valley elites have started constructing elaborate and fantastical plans for escaping the apocalypse. The imagined result of global climate change, economic catastrophe, or even left-wing uprisings against the 1 percent, the ideologies of tech neo-reactionaries – the Peter Thiels and Mencius Moldbugs of the world – turn to the persistent reactionary strategy of exit or escape to preserve a privilege they anticipate will slip from their grasp.

The Seasteading Institute's fantasy is to build an artificial island, a free-floating, hyper-libertarian city-state, bobbing on the rising oceans amid the plastic islands. Others fantasize about constructing underground bunkers in relatively "desolate" regions that have escaped extraction and depletion, such as New Zealand.[12]

Then there's the ultimate fantasy of repeating the imperial gesture on an interplanetary scale. Let's terraform Mars! That digitally accelerated economies of depletion are doing something like "Venus-forming" the earth with a runaway greenhouse effect can then be quietly forgotten.

What these white boy fantasies have in common is an escape from zones of depletion, depletion in which they may very well be implicated, before the precarity and toxicity they generate catches up with them.

The Promise of Self-Improvement

Our technocrats seek to extricate and insulate themselves from the generalized feeling of insecurity. At the same time, they have constructed networks that fueled the proliferation of digitally mediated ethno-nationalisms, open calls to abandon even the most superficial forms of democracy, published desires for a future world order comprised of hyper-reactionary, neo-mercantilist corporate monarchies, and a heaping trash of conspiracy theories about faked moon landings, vaccines causing autism, and the earth truly being flat. Culture and communication become zones depleted not just of what is factually true but of what is usefully imaginative. It is as if our digital overlords feel safer in their gated community if those of us outside of it are at each other's throats. The most provocable among those who find themselves on the wrong side of the velvet rope is the affronted class. Whiteness does not necessarily get you a first-class – or even business-class, or even economy-plus-class – seat on the lifeboat any more.

Theorists who hope to understand those who once belonged to the welfare state's synthetic white middle class are faced with a double-bind. Critique can take seriously the disaffection of a white population that feels exposed to vulnerability, many perhaps for the first time, as the conditions of precarity are becoming increasingly generalized. On the other hand, we feel nothing but disgust for the racism, gender and sexual discrimination, xenophobia, classism, and a more general hatred of difference that so often becomes the refuge from anxiety. White boys are entering the de-*pelted* zone, where their skins do not always save them. All too often their response is to strap on a layer of crusader armor.

Newly disempowered white boys also seek the restoration of their fragile egos in the promise of other forms of self-improvement – an economization and bastardization of the possibilities of covens of care. One line of self-help that seems to be specifically designed to direct white men away from more radical responses to their new-found vulnerabilities comes from high priest of the anxious white boys: Jordan Peterson.[13] With a mix of soft psychology, self-help platitudes, and pseudo-intellectual conspiracy theories about ethnic studies and women's studies, and a just-so story about lobsters, Peterson presents his followers with the gift of an object onto which they can project their nebulous anxieties: the figure of the *cultural Marxist.* He tells them that the proponents of "identity politics" are actually communists in sheep's clothing, and that postmodernism is really just masked Marxism. While capitalism further erodes their life-worlds, Peterson channels the resentment of his white male followers towards women and people of color.

Talking heads of the Peterson type exploit the dynamics of the information swamp in which we are all obliged to fish for morsels of entertainment and information. Being against something or someone in the name of a value people can be persuaded they already hold, and which is claimed as an immutable truth, generates more heat than light. The information extraction zones designed for Facebook and its rivals share the quality of feeding off quantities rather than qualities of information, and hence positive feedback loops of outrage and spite are not a bug but a valuable feature of their design.

The existence of such figures puts any counter-strategy into a double-bind. Ignore them and they capture fleeting attention, in this case of anxious white boys. Engage with them and you legitimate them and add clicks and likes to their profiles. The attempt to pull back the attention of white boys from such figures can't be imposed on all those they are so keen to other. It's a job for specialists. If one wanted an exemplary counter-tactic, one could be with Natalie Wynn, whose YouTube video channel ContraPoints began as an attempt at persuading those attracted to the language of affronted masculinity to think otherwise.[14]

What makes it work, curiously enough, is that Wynn is a trans woman. Her transition happened throughout the making of the videos, and one can trace a double-path of her providing ways of thinking counter to the mimetic rage of affronted masculinity and at the same time her moulting out of masculinity into her own person. Not every troubled white boy is a trans woman, of course. But every troubled white boy could be somebody else. Every troubled white boy has it in their power to give up on the fantasy of belonging with the Zuckerbergs and Musks on the shuttle to Mars – which will leave, if it ever does, without them anyway.

Like Wynn, what we ask of the affronted class is, instead of clutching at supremacy to numb the pain of precarity, to mobilize against engines of capitalism, which starts with an acknowledgement of shared but unevenly distributed precarity. To borrow from Fred Moten, "The coalition emerges out of your recognition that it's fucked up for you, in the same way that we've already recognized that it's fucked up for us. I don't need your help. I just need you to recognize that this shit is killing you, too, however much more softly, you stupid motherfucker, you know?"[15]

Restoring the Depleted World

We must now think and act in a depleted and depleting world: depleted by concatenating and ramifying regimes of extraction, layered and laminated on top of each other. Call it the Anthropocene or the Capitalocene or whatever you like. Some zones in this world are, indeed, far more depleted than others, and some zones still enrich themselves through the destruction of others. We are shocked by how ill-equipped we are to offer tactics to mitigate this situation. Some of us perceive power as totalizing and inevitable. Some of us feel too implicated in digital environments and their affordances to posit tactics or fixes to digital precarity. Despite the helplessness we feel, we write this chapter in an attempt to give you, our reader, something to hold on to.

It has to be acknowledged: many struggle within the struggle, without collective bargaining rights, within imminent environmental catastrophe, and atmospheric anti-blackness,[1] and in spite of the violence of police power and the criminalization of poverty and protest.

The burdens of solving precarity too often fall on those who are already the most precarious. But sometimes, one has to take two steps back to take three steps forward, and draw strength and courage from our radical forebears, our queer aunts and odd uncles who have faced these dilemmas before. We learn from woman-of-color feminist collectives and other communities of precarious people who know how to build worlds out of debris: reading groups, game nights, acts of kindness and sustenance.

Gloria Anzaldúa: "Perhaps like me you are tired of suffering and talking about suffering [...] Basta de gritar contra el viento—toda palabra es ruido si no está acompañada de acción (enough of shouting against the wind—all words are noise if not accompanied with action). Dejemos de hablar hasta que hagamos la palabra luminosa y activa (let's work not talk, let's say nothing until we've made the world luminous and active) [...] With This Bridge [...] hemos comenzado a salir de las sombras; hemos comenzado a reventar rutina y costumbres opresivas y a aventar los tabúes; hemos comenzado a acarrear con orgullo la tarea de deshelar corazones y cambiar conciencias (we have begun to come out of the shadows; we have begun to break with routines and oppressive customs and to discard taboos; we have commenced to carry with pride the task of thawing hearts and changing consciousness). Mujeres, a no dejar que el peligro del viaje y la inmensidad del territorio nos asuste—a mirar hacia adelante y a abrir paso en el monte (Women, let's not let the danger of the journey and the vastness of the territory scare us—let's look forward and open paths in these woods). Caminante, no hay puentes, se hace puentes al andar (Voyager, there are no bridges, one builds them as one walks)."[2]

We hold dominant social imaginaries responsible to the multiple and often conflicting visions of communities. These visions are rooted in histories of exploitation and domination as they imagine or design future infrastructure. In their wake, precarious subjects break the chains of enclosure for a chance to let different modes of existence circulate and propagate.

Infrastructures have long been the arteries of extraction. Digital infrastructures, from fiber optics to communications platforms, are run on dispossession, genocide, forced relocation, and extraction. Local communities have responded by "bootstrapping" digital infrastructure. Rather than owning the means of production directly, appropriating the means of *mediation* has become a way to control the lifeblood of commodity circulation. These modes of mediated extraction (re)produce racialized subjects as surplus. In what follows, we look to fragile, iterative projects that are likely to break down in some capacity. Each small failure is a cut, a fissure in the infrastructure. Highly local, non-transposable, these projects regenerate, build, and speculate other futures.

The following projects model the rebuilding of the commons after its dismantling by privatization. They do not exist fully outside of the economy of privatization but prop open enclosed infrastructures. These projects assert the community's right to control its technology. We want to hold open the extent to which these projects offer a glimpse at the possibility of constructing something like a commons that sits atop already privatized channels of communication.

However, we also recognize that projects like these may also have a more pernicious function: that of offloading the work of constructing, repairing, and maintaining infrastructures onto precarious communities that have already lost ground to the depletion economy.

Because computers offer convenience, distraction, social connection, efficiency and pleasure, we can't give them up or give up on them. These projects gesture to alternative ways of being with technology. We note, however, that our discussions of particular projects are not to be read as endorsements, nor as condemnations of organizing practices. Rather, we look to these initiatives as having a fraught orientation to precarity, capitalism, and surveillance.

Example 1: Detroit Digital Stewards Program

Post-deindustrialization efforts to "revitalize" Detroit return us to our earlier theorization of the laboratory. Detroit is often understood by white folks across political spectrums as the model of the inevitable decline of Made in America, the debris of outsourcing production. We echo the work of community leaders, activists, and black academics in refuting the characterization of Detroit-as-detritus, but call attention to the projects (both those being imposed on Detroit and those implemented by the local communities) as engaging in and/or working against the surveillant laboratory model of innovation.

We begin with the Detroit Digital Stewards Program in Detroit, Michigan, a project based in highly segregated and predominantly Latina/o/x and African-American (and, in some cases, rapidly gentrifying) neighborhoods. The Digital Stewards, supported by a partnership between Allied Media projects and Open Technology Institute, are residents from low-income neighborhoods who train local residents in installation, maintenance, and support of network technology. Some have expertise in web support or construction; others are youth or elders with no prior tech experience.

The group installs and supports mesh networks in Detroit neighborhoods, using hardware that constructs local area networks (LANs) for hyperlocal communication and shared libraries of audio and text resources. These mesh networks can also act as magnifiers for wireless internet, expanding the range of a single Wi-Fi access point to encompass not just an individual home, but an entire block. Neighbors can therefore share internet access with each other, and help each other solve connection problems along the way.

Hardware and network support are economically out-of-reach for many residents in underserved communities. The Detroit Community Technology Project, an organization that runs the Digital Stewards Program, runs a web access network for local people that also provides community self-determination. This group works with communities to decide how they will use their local area connections and monitor equitable bandwidth usage among themselves.

In recent years, this work has expanded to include the creation of a print manifesto and how-to guide on community safety in the face of digital surveillance and extraction, called "Our Data Bodies."[3]

It also includes a disaster response plan involving battery-run LAN lines, portable kits that extend the range of the network in case of breakdown, solar portable charging stations that double as maps for food and shelter in local areas, and community-organized training so that people can use the networks to facilitate resource distribution and provide shelter in the case of natural disaster.

In asserting that communication is a human right, these Detroit-based digital infrastructural projects are neither technophobic nor technophilic, neither techno-optimistic nor pessimistic. Instead, they strip "the digital" down to its fundamental affordance as a communication tool. The Detroit Community Technology Project is an experiment in social safety networking. They imagine another trajectory: what if the internet had been designed for the neighborhood, for the articulation of different socialities, rather than for the boardroom, shopping mall, and battlefield?

Example 2: Palestine and Maps.me

Israel has long experimented modes of governance and genocide on Palestinians. From limiting the number of calories per person that can enter into Gaza, to experimenting with white phosphorus, Palestinian bodies are subject to surveillance and premature death.

One of the unexpected side effects of platform capitalism is the increased precarity of subaltern subjects. For example, the supremacy of Google Maps and Alphabet-owned Waze paired with a lack of accurate maps of Palestine on these platforms has meant that Palestinian local knowledge is not integrated into these applications. Additionally, the technological apartheid enacted by Israel has meant that Palestinians cannot always access 4G networks.[4]

Therefore, a number of Palestinians have released alternative mapping applications, like QalandiaApp and Azmeh. Basel Sader, the 20-year-old law student that developed English- and Arabic-language Azmeh for iPhone and Android operating systems in 2015, explained: "This application can't give [Palestinians] the freedom of movement but it can make things easier for them."[5] These two applications feature user-submitted traffic conditions at Israeli checkpoints and are designed to run on slow networks. Within five months of its launch, 11,000 Azmeh users could find traffic data for 47 checkpoints.

In December 2017, *Wired Magazine* featured Maps.me as a step-by-step navigation application used by Palestinians that draws from open-source data and can be downloaded for offline use. The application draws on data from OpenSourceMaps (OSM), a collaborative, volunteer-run, free, and open project that creates base maps. OSM Palestine volunteers have difficulties mapping Palestine as well, like constantly changing geographical and traffic conditions due to Israeli bombings and checkpoints.[6] Furthermore, most Palestine mappers are not locals – they tend to be Israelis of all political orientations, or humanitarian activists using satellite data. Mappers discuss how to distinguish information from noise.

> “there are some tracks that go over landfill, farmyards, roads and buildings that seem to be tank traces, im taging [sic] them as ~~|highway=track | tracktype=grade3 | note=seems to be tank tracks |~~”
>
> “I believe we should only be showing permanent features and that tank tracks are not permanent. Shell holes might be if they are big enough (many became fishing ponds in Vietnam) but the ones in Gaza don’t seem to be that big.”[7]

QalandiaApp, Azmeh, OSM, and Maps.me gesture to questions of whether violence creates geographic features, questions around access and accountability, of what information can or should be mapped, who should be mapping it, and for what purpose. Rather than understanding maps as mundane objects, these projects indicate the stakes of application-based mapping tools. Maps can be crucial for survival, not just for finding the nearest Starbucks.

Privatized systems prey on racial minorities; these regenerating economies allow for experimenting with new forms of life and solidarity. Gloria Anzaldúa believed that radical politics need to be grounded in the body, in aesthetic creativity, in community, "there are no bridges, one builds one as one walks."

What you do is always tenuous, in process and built in compromised terrain. As a result, building as you go requires a different way of articulating the commons. Echoing the Zapatista proverb, "asking, we walk" (*preguntando caminamos*).

You, we, us need to allow the process of questioning what is to be done to inform what we do and to bring us together in difference as a way to produce collectivity. This is what the digital projects in Detroit and Palestine have been doing. This is what the process of writing this book has been. In the exercise of questioning, we moved forwards and sideways, though more importantly we momentarily loosened the turn of the screw.

Precarious lives bear the burden of risk and uncertainty that are the residual effects of our extractive and depleting post-digital economy. Collectivist projects and regenerating economies inject hope and fear in connections from old broken worlds to new worlds of endurance, one tenuous thread at a time.

Under what can feel like an unbearable present, some have the strength to build anew. Others retreat, seeking tender refuge in the undercommons of networks of alternative kinship, in covens of care.

Covens of Care

Bubble, bubble, toil and trouble: covens of care exist in a state of stubborn feminist killjoy hopefulness.

Covens are usually thought of as gatherings of witches, and maybe they still are, although in the Middle Ages monks could coven as well. The ancient witches' coven was generally thought to have "had 13 members: six men and six women plus a high priestess" to "produce the best harmony and results in magic."[1] This was not a rigid rule, but size was something to be considered as "too few members mean[t] ineffective magic [and] [t]oo many became unwieldy."[2] The role of the high priestess was more a function of psychic intuition and dedication to the administration of the coven, rather than a reproduction of the oppressive, power-hungry electorate that by covening the witches had desired to escape.

Thus, maybe there's something a little witchy, a little clandestine, a little magical, a little queer, about the forms of communal care and convocation that stubbornly endure in and against regimes of network governance and value extraction. Covens have familiars, but are not necessarily family. Covens have covenants, but not binding contracts. To networks of value extraction, covens are an ever-receding hinterland that is never quite entirely tracked and monetized. They sit in the gaps and fissures of the logics of accumulation and anticipation.

From the point of view of the coven itself, it is the coven that makes a place, a center, a hearth in a heartless world. Covens of care endure despite conditions of domination, violence, or erasure. At times, covens rely on digital interfaces and networks. They form and endure both in and against the precarious conditions of life that networks impose. In the connected world, real or imagined, being physically present with desired others becomes a privilege enjoyed by the few; for example, while transnational labor markets move Filipina nannies abroad, they might only communicate with their own children back home via WhatsApp.

Remember: you are sleeping for the boss! All of one's time, one's cognitive powers, one's emotional strength, is supposed to be on the job, and can be called to work at any time. Google's headquarters in New York City has "themed" meeting rooms, such as the one that looks like a baroque antechamber as reimagined by Florida real estate agents. One of the rooms attempts to simulate the living room of an average middle-class, middle-American suburban family, with its coffee table, plush sofa and table lamps. There is even a clutter of children's toys in the corner. One can only imagine the feelings of actual employees, who have families, who have kids, called into weekend meetings in this room that takes pains to duplicate all they are obliged to be away from. Even if one has a heteronormative family or something like it to turn back to, one might still need a coven of care to pick up the pieces.

The ever-presence of information and connection carves individual time into seemingly random and even chaotic obligations to wage labor. Rather than free the worker from the tyranny of the workplace, the cellphone means the workplace is everywhere. The whole of space becomes an open-plan office, in which shouty men yabber into their phones as if the fate of the world depended on it. The coven, in response, hides in plain sight, in a group message, in a Slack backchannel, for those who don't get to pretend they are masters of the universe.

We take inspiration from a graduate student feminist coven at the University of Kentucky. The coven communes in person but also through private messages and group chat, sharing support through empowering hashtags or memes that critique their condition. They recognize the university as a space of "competition, scarcity, imperialism, racism, and patriarchy," but their shared aim is not so lofty and exhausting as trying to tackle these cultural structures of precarity all at once.[3] Instead, their goal is to create autonomous feminist zones of empowerment and support that exist in and survive outside of the university. The coven tackles its own precarity through self-care and community, welcoming care while acknowledging all the bullshit happening that has it needing it.

> Instead of competing with each other, we collaborate by praising boldness, cultivating norms of trust rather than suspicion, elevating friendship above romance, grounding our relationships within political work and feminist praxis. Our willfulness to love and resist conjures momentary, inhabitable spaces, where we dream of alternative futures and nurture our energies for revolutionary change. The autonomous feminist spaces we create – these groups, our friendships – give us the strength and the enthusiasm not to settle for the few comforts of professionalization in the university. We know that the success of some comes on the backs of more precarious others. We acknowledge the cruel optimism of holding onto dreams of recognition and respect in the academy. We are in an unhopeful condition, a kind of catastrophe or impasse, and we stay here anyway.[4]

The covens of care through which people connect and support each other may have long-lived roots that extend deep into the ground of a given place. Or, they may assemble briefly and disperse, like cloud formations in virtual space. Many people are forced into mobility even as the fantasy of the digital says that mobility is not necessary. As the dictates of work make people more and more mobile, uprooting themselves and implanting themselves in city after city, online covens of care can, at times, become the only constant.

Covens of care take more than a casual commitment. They mean putting something else before the dictates of labor, seeing them as being of more value, even if their instantiation can only happen in the margins of work's demands. And yet covens of care do not just persist despite these conditions, but can also produce the conditions for the possibility of living otherwise. Back in the 1970s, Angela Y. Davis described the domestic care work of women under slavery in the US, which was performed not in a family household, but for men and children and other women who came together not as kin but as kith under conditions of complete domination.[5] This other kind of domestic space became the main space of resistance to the conditions of slavery, because the care labor within it was the only work not fully claimed by the slave owner. While reproducing the lives of the enslaved as property, this care work also created the conditions for resistance to enslavement.

A different kind of example: consider a network of trans women, mostly living in New York City. A few have straight jobs. Some do sex work. Some are casually employed here and there. Sometimes they live together. Sometimes they fuck each other. Sometimes they gossip and snipe. And yet, always, there's the watching out for sisters. Someone has surgery and needs others to come care for her. Someone is suicidal and someone has to look out for her. Someone needs to borrow a shot of estrogen as their prescription ran out. Someone is organizing a benefit to raise money for someone else's surgery. Everyone puts ten dollars in each other's GoFundMe for emergency rent money or medical expenses. Everyone posts selfies with questionable new outfits for others to affirm or critique.

It's a coven of care. It is not utopia. This person is no longer on speaking terms with that one because of that one thing she did - and so on and so forth. Indeed, covening is always fraught, for it involves the mobilization of social networks of care that are only available to those who are already stitched to people with resources -emotional and material. Covens are often organized into racial, subcultural, and class-segregated units. The Jane Collective was a kind of coven consisting mostly of white women.[6] Queer land projects in Tennessee have historically provided refuge for mostly white people who invented new modes of care in the wake of the AIDS crisis.[7] While covens are a crucial mode of recirculating resources of care, not everyone can, or knows how to, access the coven. Shy radicals, working-class women of color, and those most worn-down by precarity may have no way (and no energy) to enter the coven.

The coven, almost by definition, is unstable – for covens emerge in and against spaces of institutionalization. Their ongoingness is always in peril. People who need constant care may not find it in the coven. Survivors of abuse who only have their coven may quickly find they need more buoying-up than their friends can provide.

But when most of the world hates you or is at best indifferent to you, the coven is a place from which to draw strength, and maybe even to invent a new way of life. Some have called this way of life T4T: trans for trans.[8] Sometimes, all we have is each other.

Something that is obvious to trans people maybe sometimes more than to others is that flesh and tech are integral, for everyone, but in such extremely variable ways. Trans people can need hormones, surgeries, interventions of tech into flesh just to make life livable. But perhaps all bodies are like this, including cis bodies. Bodies are at once more and more precarious and more and more dependent on systems of technique to function.[9] Covens of care create pockets of shared affect and attention, in and against medical-technical business models that rank and rate bodies mostly on their profitability and ability to pay.

What does covening in care feel like? José Esteban Muñoz gave us "feeling Brown" as a way to be in the white supremacist nation but not of it, a connection without identity, and for some a connection with other subaltern histories that cannot be spoken.[10] For people of color and other precarious subjects who feel not quite right, Muñoz soothes us: feeling Brown is a way to feel one another, together. It can feel excessive, overperformed. It can feel like a strange affinity. But it feels less precarious, together, now.

Through covening, we preserve the incalculable – the abundance of "the gift" and its attendant obligations – to give, receive, and reciprocate.[11] The gift puts you in a relation of reciprocity, where one's ongoing debt sustains a relationship. It takes a lot of effort to take a gift *out* of a commodity. We create emergent and messy zones of intimacy that exceed monetized transactions. The coven emerges, not necessarily as a romantic fuck-you to the disciplinary function of welfare state, but in the crucible of mass abandonment – by both the state and legitimized networks of social support. When you are kicked out of the family, you lose access to society's default network of care. This is why covens emerge as complementary networks for the affective value of the family. We coven because our survival depends on it.

In an ever-more-depleted world, covens have to bear a supplementary burden of finding ways to sustain life outside of commodified enrichment. They come under pressure to be what gets us through the superadded volatility not just of the social-technical world but of the natural-cultural one as well. Maybe it's a time to learn from – and teach the wisdom of – various kinds of witches.

Notes

The Precarity Effect: On the Digital Depletion Economy

1 Justin Joque, *Infidel Mathematics* (New York: Verso, forthcoming).
2 See Seb Franklin; digitality can be understood "not only as a logical-technical substrate through which certain machines might operate but also as a predominant logical mode with which to address both individual social actors and the body of interactions between these actors that can be dubbed 'society.'" Seb Franklin, *Control: Digitality as Cultural Logic* (Cambridge, MA: The MIT Press, 2015), xviii.
3 Ruth Wilson Gilmore, *Golden Gulag: Prisons, Surplus, Crisis, and Opposition in Globalizing California* (Berkeley, CA: University of California Press, 2007), 28.
4 Lauren Berlant, *Cruel Optimism* (Durham, NC: Duke University Press, 2011).
5 Precarity Lab, "Digital Precarity Manifesto," *Social Text* 37, no. 4 (2019): 77–93.
6 Shoshana Zuboff, *The Age of Surveillance Capitalism: The Fight for a Human Future at the New Frontier of Power* (New York: Public Affairs, 2019).
7 Nick Srnicek, *Platform Capitalism* (Malden, MA: Polity, 2017).
8 Gilles Deleuze, "Postscript on the Societies of Control," *October* 59 (1992): 3–7.
9 Karl Marx, *Capital, Volume I: A Critique of Political Economy*, trans. Ben Fowkes (New York: Penguin, 1976).
10 Cedric Robinson, *Black Marxism: The Making of the Black Radical Tradition* (Durham, NC: Duke University Press, 1993).
11 Guy Standing, *The Precariat: The New Dangerous Class* (New York: Bloomsbury, 2011).
12 John Button, *A Dictionary of Green Ideas: Vocabulary for a Sane and Sustainable Future* (New York: Routledge, 1998).
13 Anna Lowenhaupt Tsing, *The Mushroom at the End of the World: On the Possibility of Life in Capitalist Ruins* (Princeton, NJ: Princeton University Press, 2015).
14 Stefano Harney and Fred Moten, *The Undercommons: Fugitive Planning & Black Study* (Brooklyn, NY: Autonomedia, 2013).

15 Donna Haraway, “Situated Knowledges: The Science Question in Feminism and the Privilege of Partial Perspective,” *Feminist Studies* 14, no. 3 (1988): 575–599.

16 Geof Bowker, “How to be Universal: Some Cybernetic Strategies, 1943–1970,” *Social Studies of Science* 23, no. 1 (1993): 107–127.

17 Max Weber, *Economy and Society: A New Translation*, ed. and trans. Keith Tribe (Cambridge, MA: Harvard University Press, 2019); Michel Foucault, *The Order of Things* (New York: Routledge, 2002).

18 See, for instance: Bruno Latour and Steve Woolgar, *Laboratory Life: The Construction of Scientific Facts*, 2nd edn (Princeton, NJ: Princeton University Press, 1986); Steven Shapin and Simon Schaffer, *Leviathan and the Air-Pump: Hobbes, Boyle, and the Experimental Life* (Princeton, NJ: Princeton University Press, 2018); Donna Haraway, *Modest_Witness@Second_Millennium.FemaleMan©_Meets_OncoMouse™: Feminism and Technoscience* (New York: Routledge, 1997); Karin Knorr Cetina, *Epistemic Cultures: How the Sciences Make Knowledge* (Cambridge, MA: Harvard University Press, 1999).

19 Mary Jo Deegan, *Race, Hull-House, And The University Of Chicago: A New Conscience Against Ancient Evils* (Westport, CT: Praeger, 2002); David E. Apter, Herbert J. Gans, Ruth Horowitz, Gerald D. Jaynes, William Kornblum, James F. Short, Gerald D. Suttles, and Robert E. Washington, “The Chicago School and the Roots of Urban Ethnography: An Intergenerational Conversation with Gerald D. Jaynes, David E. Apter, Herbert J. Gans, William Kornblum, Ruth Horowitz, James F. Short Jr., Gerald D. Suttles, and Robert E. Washington,” *Ethnography* 10, no. 4 (2009): 375–396.

20 Arturo Escobar, *Encountering Development: The Making and Unmaking of the Third World* (Princeton, NJ: Princeton University Press, 1994), 36.

21 Michelle Murphy, *The Economization of Life* (Durham, NC: Duke University Press, 2017).

22 Fouzieyha Towghi and Kalindi Vora, “Bodies, Markets, and Experiments in South Asia,” *Ethnos: Journal of Anthropology* 70, no. 1 (2014): 1–18.

23 Hortense Spillers, “Mama’s Baby, Papa’s Maybe: An American Grammar Book,” *Diacritics* 17, no. 2 (1987): 64–81.

24 Ian Baucom, *Specters of the Atlantic: Finance, Capital, Slavery, and the Philosophy of History* (Durham, NC: Duke University Press, 2005).

25 Walter Johnson, *River of Dark Dreams: Slavery and Empire in the Cotton Kingdom* (Cambridge, MA: Harvard University Press, 2017), 87.

26 Steven Beckert, *Empire of Cotton: A Global History* (New York: Vintage Press, 2014), xvii.

27 The Trans-Canada Pipeline Pavillion is a named building at Banff, a few steps away from where we write. The building of a Trans Mountain line through Jasper National Park, just to the North of Banff, which some Albertan politicians hoped would fund social services programs in the province, is still controversial. Darryl Dyck, "Trans Mountain, Trudeau and First Nations," *The Globe and Mail*, April 27, 2018, www.theglobeandmail.com/politics/article-trans-mountain-kinder-morgan-pipeline-bc-alberta-explainer/.

We mention Canada here because we write from this location during this time, but all of us are from elsewhere, places with their own histories of settler-colonialism and imperial legacy. See our bios for more on our individual locations. No shade on Canada in particular.

28 LeAnn Howe, "The Chaos of Angels," *Callaloo* 17, no. 1 (1994): 108.

29 Jodi Byrd, *Transit of Empire: Indigenous Critiques of Colonialism* (Minneapolis, MN: University of Minnesota Press, 2011), xxviii.

30 Saidiya Hartman, "The Anarchy of Colored Girls Assembled in a Riotous Manner," *South Atlantic Quarterly* 117, no. 3 (2018): 465–490.

31 Catherine Knight Steele, "Signifyin' Bitching and Blogging: Black Women and Resistance Discourse Online," in *The Intersectional Internet: Race, Sex, Class and Culture Online*, eds Safiya Umoja Noble and Brendesha M. Tynes (New York: Peter Lang, 2016), 73–93.

32 Jade Y. Power-Sotomayor, "From *Soberao* to Stage: Afro-Puerto Rican and the Speaking Body," in *The Oxford Handbook of Dance and Theater*, ed. Nadine George-Graves (Oxford: Oxford University Press, 2015).

The Undergig

1 Aihwa Ong, *Neoliberalism as Exception: Mutations in Citizenship and Sovereignty* (Durham, NC: Duke University Press, 2006).

2 Mario Tronti, *Workers and Capital*, trans. David Broder (New York: Verso, 2019); Antonio Negri, *The Politics of Subversion: A Manifesto for the Twenty-First Century* (Malden, MA: Polity, 2005).

3 Sylvia Wynter, "Unsettling the Coloniality of Being/Power/Truth/Freedom: Towards the Human, After Man Its Overrepresentation–An Argument," *CR: The New Centennial Review* 3, no. 3 (2003): 257–337.

4 Raj Patel and Jason W. Moore, *A History of the World in Seven Cheap Things: A Guide to Capitalism, Nature, and the Future of the Planet* (Oakland, CA: University of California Press, 2017).

5 Achille Mbembe, *Critique of Black Reason*, trans. Laurent Dubois (Durham, NC: Duke University Press, 2017), 17–18.

6 Sandro Mezzadra and Brett Neilson, *The Politics of Operations: Excavating Contemporary Capitalism* (Durham, NC: Duke University Press, 2019).

7 Ann Laura Stoler, *Duress: Imperial Durability in Our Times* (Durham, NC: Duke University Press, 2016).

8 *Ibid.*, 8.

9 Os Keyes, Nikki Stevens and Jacqueline Wernimont, “The Government Is Using the Most Vulnerable People to Test Facial Recognition Software,” *Slate*, March 17, 2019, https://slate.com/technology/2019/03/facial-recognition-nist-verification-testing-data-sets-children-immigrants-consent.html.

10 Sean Hollister, “Google Contractors Reportedly Targeted Homeless People for Pixel 4 Facial Recognition,” *The Verge*, October 2, 2019, www.theverge.com/2019/10/2/20896181/google-contractor-reportedly-targeted-homeless-people-for-pixel-4-facial-recognition.

11 Kalindi Vora, *Life Support: Biocapital and the New History of Outsourced Labor* (Minneapolis, MN: University of Minnesota Press, 2015).

12 Lily Irani, “The Cultural Work of Microwork,” *New Media & Society* 17, no. 5 (2015): 720–739.

13 Mezzadra and Neilson, *The Politics of Operations*, 100.

14 Dipesh Chakrabarty, *Provincializing Europe: Postcolonial Thought and Historical Difference* (Princeton, NJ: Princeton University Press, 2000).

15 Berlant, *Cruel Optimism*.

16 Harney and Moten, *The Undercommons*.

Techno Toxic

1 Mel Chen, “Unpacking Intoxication, Racialising Disability,” *Medical Humanities* 41, no. 1 (2015): 25–29.

2 Brett Walker, *Toxic Archipelago: A History of Industrial Disease in Japan* (Seattle, WA: University of Washington Press, 2010).

3 Alexis Shotwell, *Against Purity: Living Ethically in Compromised Times* (Minneapolis, MN: University of Minnesota Press, 2016), 6–7.

4 Andrew Ross, *Bird on Fire: Lessons from the World’s Least Sustainable City* (Oxford: Oxford University Press, 2011).

5 Adrian Chen, “The Laborers Who Keep Dick Pics and Beheadings Out of Your Facebook Feed,” *Wired*, October 23, 2014, www.wired.com/2014/10/content-moderation/; Ciaran Casidy and Adrian Chen, “The Moderators,” *Field of Vision*, April 2017, https://vimeo.com/213152344.

6 Mary L. Gray and Siddharth Suri, *Ghost Work: How to Stop Silicon Valley from Building a New Global Underclass* (Boston, MA: Houghton Mifflin Harcourt, 2019).
7 Mitali Thakor, "Digital Apprehensions: Policing, Child Pornography, and the Algorithmic Management of Innocence," *Catalyst: Feminism, Theory, Technoscience* 4, no. 1: 1–16.
8 *Ibid.*
9 L.M. Shields, W.H. Wiese, B.J. Skipper, B. Charley, and L. Benally, "Navajo Birth Outcomes in the Shiprock Uranium Mining Area," *Health Physics* 63, no. 5 (1992): 542.
10 Catherine Sue Ramírez, "Deus Ex Machina: Tradition, Technology, and the Chicanafuturist Art of Marion C. Martinez," *Aztlán* 29, no. 2 (2004): 55–92.
11 Cam Simpson, "American Chipmakers Had a Toxic Problem. Then They Outsourced It," *Bloomberg Businessweek*, June 15, 2017.
12 *Ibid.*
13 Mel Y. Chen, "Toxic Animacies, Inanimate Affections," *GLQ: A Journal of Lesbian and Gay Studies* 17, nos 2–3 (2011): 265–286.

The Widening Gyre of Precarity

1 Dan Georgakas and Marvin Surkin, *Detroit, I Do Mind Dying: A Study in Urban Revolution* (New York: Southend Press, 1998).
2 Wendy Hui Kyong Chun, *Control and Freedom: Power and Paranoia in the Age of Fiber Optics* (Cambridge, MA: The MIT Press, 2008), 28.
3 William Butler Yeats, "The Second Coming," in *The Collected Works of W.B. Yeats*, 2nd edn, ed. Richard J. Finneran (New York: Scribner Paperback Poetry, 1996), 187.
4 Samuel Beckett, *Samuel Beckett's Molloy, Malone Dies, The Unnamable*, ed. Harold Bloom (New York: Chelsea House Publishers, 1998).

Automating Abandonment

1 Bureaucrats create formal rules and procedures for determining who is eligible or ineligible for public assistance programs and computers execute them. Virginia Eubanks, *Automating Inequality: How High-Tech Tools Profile, Police, and Punish the Poor* (New York: St. Martin's Press, 2017).
2 H. Claire Brown, "How an Algorithm Kicks Small Businesses Out of the Food Stamps Program on Dubious Fraud Charges," *The Counter*, October 8, 2018,

https://thecounter.org/usda-algorithm-food-stamp-snap-fraud-small-businesses/.

3 Zach Friedman, "99% Rejected for Student Loan Forgiveness – Again," *Forbes*, September 9, 2019, www.forbes.com/sites/zackfriedman/2019/09/09/99-rejected-for-student-loan-forgiveness-again/#69a05d1d6675.

4 Eubanks, *Automating Inequality*.

Fantasies of Ability

1 Berlant, Cruel Optimism.

2 Rob Nixon, *Slow Violence and the Environmentalism of the Poor* (Cambridge, MA: Harvard University Press, 2011).

3 Rae Ellen Bichell, "Scientists Start to Tease out the Subtler Ways Racism Hurts Health," *NPR*, November 11, 2017, www.npr.org/sections/health-shots/2017/11/11/562623815/scientists-start-to-tease-out-the-subtler-ways-racism-hurts-health.

4 Sarah F. Rose, *No Right to Be Idle: The Invention of Disability, 1840s–1930s* (Chapel Hill, NC: University of North Carolina Press, 2017).

5 Saidiya Hartman, *Wayward Lives, Beautiful Experiments: Intimate Histories of Social Upheaval* (New York: W.W. Norton & Company, Inc., 2019).

6 Muriel Rukeyser, *The Book of the Dead* (Morgantown, WV: University of West Virginia Press, 2018).

7 Adelina Lancianese, "Before Black Lung, the Hawk Nest Disaster Killed Hundreds," *NPR*, January 20, 2019, www.npr.org/2019/01/20/685821214/before-black-lung-the-hawks-nest-tunnel-disaster-killed-hundreds.

8 Rose, *No Right to Be Idle*.

9 For more on medical technologies and global circuits of care, see Aren Z. Aizura, *Mobile Subjects: Transnational Imaginaries of Gender Reassignment* (Durham, NC: Duke University Press, 2018) and Jules Gill-Peterson, "The Techical Capacities of the Body: Assembling Race, Technology, and Transgender," *Transgender Studies Quarterly* 1, no. 3 (2014): 402–418.

10 Krissy Clark, "The Disconnected," *Slate*, January 3, 2016, https://slate.com/news-and-politics/2016/06/welfare-to-work-resulted-in-neither-welfare-nor-work-for-many-americans.html; Rachel Garfield, Robin Rudowitz, and Kendal Orgera, "Understanding the Intersection of Medicaid and Work: What Does the Data Say?" *Kaiser Family Foundation*, August 8, 2019, www.kff.org/medicaid/issue-brief/understanding-the-intersection-of-medicaid-and-work-what-does-the-data-say/.

11 Elizabeth Warren and Amelia Warren Tyagi, *The Two-Income Trap: Why Middle-Class Parents are Going Broke* (New York: Basic Books, 2003).
12 Frano "Bifo" Berardi, *The Soul at Work: From Alienation to Autonomy*, trans. Francesca Cadel and Giuseppina Mecchia (Los Angeles, CA: Semiotext(e), 2009).
13 Arlie Russel Hochschild, "Emotion Work, Feeling Rules, and Social Structure," *American Journal of Sociology* 85, no. 3 (1979): 551–575.
14 Max Baumkel, "The Invisible Presence of Trans-Bodies: Unpacking Regimes of Visibility and Visuality Through Tom Cho's *Look Who's Morphing*," Thesis, Vanderbilt University, 2015.
15 Jennifer Proctor, "A Failure of Imagination: The Role of Disability in *Avatar*," *Media Commons*, August 10, 2010, http://mediacommons.org/imr/2010/08/03/failure-imagination-role-disability-avatar; Rosie Forbes, "What 'Detective Pikachu' Got Wrong About Disability," *The Mighty*, May 17, 2019, https://themighty.com/2019/05/detective-pikachu-disability-villain/?utm_source=pin_board_disability&utm_medium=pinterest&utm_campaign=pin_disability_2019week18.
16 For more on cure and ableist fantasy, refer to Eli Clare, *Brilliant Imperfection: Grappling with Cure* (Durham, NC: Duke University Press, 2017).
17 On the concept of the "social factory," see Mario Tronti, *Workers and Capital*, trans. David Broder (New York: Verso, 2019); Mario Tronti, "Factory and Society," *Operaismo in English*, June 13, 2013, https://operaismoinenglish.wordpress.com/2013/06/13/factory-and-society/; Antonio Negri, *Marx Beyond Marx: Lessons on the Grundrisse*, trans. Harry Cleaver (New York: Autonomedia, 1992). For a more contemporary discussion, see Trebor Scholz, ed., *Digital Labor: The Internet as Playground and Factory* (New York: Routledge, 2013).
18 Cassius Adair, "White Trans Politics and the Early Internet," paper presented at the American Studies Association Annual Meeting, Atlanta, Georgia, November 8–11, 2018.

Dispossession by Surveillance

1 Judith Butler, *Undoing Gender* (New York: Routledge, 2004).
2 Iván Chaar-López, "Sensing Intruders: Race and the Automation of Border Patrol," *American Quarterly* 71, no. 2 (2019): 495–518.
3 Zuboff, *The Age of Surveillance Capitalism*.

4 In our treatment of the plantation and the factory as laboratories we showed how, at its core, capitalism requires surveillance. See also E.P. Thompson, "Time, Work Discipline, and Industrial Capitalism," *Past & Present* 38 (1967): 56–97; Anthony Giddens, *The Nation-State and Violence: Volume Two of A Contemporary Critique of Historical Materialism* (Cambridge: Polity, 1985). The expansion of markets entails observation and analysis of consumer habits and laborers, and primitive accumulation across the American hemisphere has always entailed the surveillance of indigenous populations.
5 Kashmir Hill, "The Secretive Company That Might End Privacy as We Know It," *The New York Times*, January 18, 2020, www.nytimes.com/2020/01/18/technology/clearview-privacy-facial-recognition.html.
6 Library of Congress, "Fugitive Slave Ads: Topics in Chronicling America," Library of Congress Research Guides, https://guides.loc.gov/chronicling-america-fugitive-slave-ads.
7 Angela Y. Davis – who was part of a 2011 delegation of women of color who traveled to Palestine – has also examined the carceral infrastructural links between Israel, Europe, and the US in her analysis of G4S, a security corporation that operates globally: "G4S has insinuated itself into our lives under the guise of security and the security state – from the Palestinian experience of political incarceration and torture to racist technologies of separation and apartheid; from the wall in Israel to prison-like schools in the US and the wall along the US-Mexico border." Angela Y. Davis, *Freedom is a Constant Struggle: Ferguson, Palestine, and the Foundations of a Movement* (Chicago, IL: Haymarket Books, 2016). See also Jewish Voices for Peace, "Deadly Exchange: The Dangerous Consequences of American Law Enforcement Trainings in Israel," 2018, https://deadlyexchange.org/wp-content/uploads/2019/07/Deadly-Exchange-Report.pdf.
8 Meryem Kamil, "Towards Decolonial Futures: New Media, Digital Infrastructures, and Imagined Geographies of Palestine," Dissertation, University of Michigan, 2019.
9 Meryem Kamil, "Post Spatial, Post Colonial: Accessing Palestine in the Digital," *Social Text* 38, no. 3 (114) (2020): 55–82.
10 Alfred W. McCoy, *Policing America's Empire: The United States, the Philippines, and the Rise of the Surveillance State* (Madison, WI: The University of Wisconsin Press, 2009), 294.
11 Micah Houser, "The High Price of Freedom for Migrants in Detention," *The New Yorker*, March 12, 2019, www.newyorker.com/news/news-desk/the-high-price-of-freedom-for-migrants-in-detention.

12 Ruha Benjamin, *Race After Technology: Abolitionist Tools for the New Jim Code* (Medford, MA: Polity, 2019).
13 Chaar-López, "Sensing Intruders."
14 Michael E. Miller, "This Company is Making Millions from America's Broken Immigration System," *The Washington Post*, March 9, 2017, www.washingtonpost.com/local/this-company-is-making-millions-from-americas-broken-immigration-system/2017/03/08/43abce9e-f881-11e6-be05-1a3817ac21a5_story.html.
15 Simone Browne, *Dark Matters: On the Surveillance of Blackness* (Durham, NC: Duke University Press, 2015).

The Affronted Class

1 Alexander Saxton, *The Indispensable Enemy: Labor and the Anti-Chinese Movement in California* (Berkeley, CA: University of California Press, 1971).
2 Silvia Lindtner, *Prototype Nation: China and the Contested Promise of Innovation* (Princeton, NJ: Princeton University Press, 2020).
3 Joque, *Infidel Mathematics.*
4 McKenzie Wark, *A Hacker Manifesto* (Cambridge, MA: Harvard University Press, 2004).
5 Lindtner, *Prototype Nation.*
6 Judy Wajcman, *Feminism Confronts Technology* (University Park, PA: Pennsylvania State University Press, 1991).
7 Chris Anderson, *Makers: The New Industrial Revolution* (New York: Crown Business, 2012).
8 *Ibid.*
9 This is what Lindtner calls the "socialist pitch" in *Prototype Nation*, by merging it with the advances of digital fabrication and automation.
10 Mark Zuckerberg, "Opening Statement to the Senate Judiciary and Commerce Committees on Facebook Data Privacy," *American Rhetoric*, April 10, 2018, www.americanrhetoric.com/speeches/markzuckerbergcongressopeningstmt.htm.
11 Erik Hersman, "Reporting Crisis via Texting," TED video, 3:44, filmed November 2009, www.ted.com/talks/erik_hersman_reporting_crisis_via_texting/details?language=bi.
12 Matthew Gault, "Hunting Silicon Valley's Doomsday Bunkers," *Vice*, May 20, 2019, www.vice.com/en_us/article/kzmzyx/hunting-silicon-valleys-doomsday-bunkers-in-new-zealand-documentary.

13 Dorian Lynskey, "How Dangerous is Jordan B. Peterson, The Rightwing Professor Who 'Hit a Hornets' Nest?" *The Guardian*, February 7, 2018, www.theguardian.com/science/2018/feb/07/how-dangerous-is-jordan-b-peterson-the-rightwing-professor-who-hit-a-hornets-nest.
14 Natalie Wynn, ContraPoints, www.contrapoints.com.
15 Harney and Moten, *The Undercommons*.

Restoring the Depleted World

1 Christina Sharpe, *In the Wake: On Blackness and Being* (Durham, NC: Duke University Press, 2016).
2 Gloria Anzaldúa, "Foreword," in *This Bridge Called My Back: Writings by Radical Women of Color*, eds Cherríe Morraga and Gloria Anzaldúa (New York: Kitchen Table: Women of Color Press, 1983), iii–iv.
3 Allied Media Projects, *Our Data Bodies: Digital Defense Playbook* (Detroit, MI: Allied Media Projects, 2019).
4 Helga Tawil-Souri and Miriyam Aouragh, "Intifada 3.0? Cyber Colonialism and Palestinian Resistance," *The Arab Studies Journal* 22, no. 1 (2014): 102–133.
5 Areej Hazboun, "New Apps Help Palestinians Navigate Israeli Checkpoints," *AP News*, November 18, 2015, https://apnews.com/117d524e2aba4ab19ec01c3d3e50aa06.
6 Kamil, "Towards Decolonial Futures."
7 *Ibid*.

Covens of Care

1 Rosemary Guiley, *The Encyclopedia of Witches, Witchcraft and Wicca* (New York: Facts On File, 2008), 77.
2 *Ibid*.
3 Araby Smyth, Jess Linz, and Lauren Hudson, "A Feminist Coven in the University," *Gender, Place and Culture: A Journal of Feminist Geography* (2019): 2.
4 *Ibid*., 3.
5 Angela Y. Davis, "Reflections on the Black Woman's Role in the Community of Slaves," *The Black Scholar: Journal of Black Studies and Research* 4 (1971): 2–15.
6 Rainey Horwitz, "The Jane Collective (1969–1973)," *The Embryo Project Encyclopedia*, August 7, 2017, https://embryo.asu.edu/pages/jane-collective-1969-1973.

7 Idyll Dandy Arts, "About Ida," https://idylldandyarts.tumblr.com/about.

8 Torrey Peters, *Infect Your Friends and Loved Ones* (CreateSpace Independent Publishing Platform, 2016).

9 Paul B. Préciado, *Testo Junkie: Sex, Drugs, and Biopolitics in The Pharmacopornographic Era* (New York: The Feminist Press, 2013).

10 José Esteban Muñoz, "Feeling Brown, Feeling Down: Latina Affect, the Performativity of Race, and the Depressive Position," *Signs* 31, no. 3 (2006): 675–688.

11 Marcel Mauss, *The Gift: Forms and Functions of Exchange in Archaic Societies* (New York: Norton, 1967).

Glossary

Affronted class A class of workers whose jobs and social status have been made newly vulnerable by automation. We call them the "affronted class" because their authority and authorship of tech innovation has been compromised. Their previously secure positions in the technology and creative sectors protected them from occupational precarity, and they are enduring a painful period of cultural adjustment to it. (See Chapter 8: The Affronted Class.)

Covens of care Covens of care endure despite conditions of domination, violence, or erasure. "Covens have familiars, but are not necessarily family. Covens have covenants, but not binding contracts. From the point of view of networks of value extraction, they are an ever-receding hinterland that is never quite entirely tracked and monetized. They sit in the gaps and fissures of the logics of accumulation and anticipation." (See Chapter 10: Covens of Care.)

Depletion zone Depletion zones are inhabited by already precarious peoples such as racial, ethnic, and sexual minorities, women, indigenous people, and migrants. They are critical sites where raw materials, labor, and information are extracted and rarely replaced. These zones are laboratories or experimental zones where precarity generates possible techniques of control that, if they "work," become generalized. (See Chapter 1: Precarity Lab.)

Depletion economy An economic system that requires the material, psychic, bodily, spiritual, and social depletion of land, environment, animals, and people. Depletion economies export toxicity and precarity to sustain the enriched world.

Digital precarity Individuals, communities, environments, and zones that contribute the raw and processed material for digital technology become depleted and insecure in their bodies and lifeworlds. Technoculture produces toxic materials, behaviors, and economies that are exacerbated by the use of digital platforms, Big Tech's capital, and the gig and undergig economies that enrich them.

Enriched zones Spaces whose resources, labor, and raw materials depend upon the extraction of "cheap nature" and culture from depleted zones. These spaces, such as Silicon Valley, Seattle, New York City, London, and Shanghai, are often defined by thriving technology industries that deplete resources from within their territorial borders.

Experimentation "The process of conducting tests that brings together the experimenter/researcher and the study subject/object" (Fouzieyha Towghi and Kalindi Vora, "Bodies, Markets, and Experiments in South Asia," *Ethnos: Journal of Anthropology* 70, no. 1 (2014): 1–18). We discuss experimentation to think beyond methods of testing, but rather to think of resource-intensive "innovation" as a form of power that fractures populations into investigators and test subjects, into extractable and governable matter. (See Chapter 1: Precarity Lab.)

Extractive zones The word "extraction" emphasizes that a resource was *taken,* not given. (See also: Macarena Gómez-Barris, *The Extractive Zone: Social Ecologies and Decolonial Perspectives* (Durham, NC: Duke University Press, 2017)). When areas, land, regions, jurisdictions, and districts are emptied, with varying degrees of force and coercion, of resources, expertise, skills, people, and lives, we know we are in an extractive zone. In scenarios where something abstract is "extracted" – such as value from data – life is reduced to a resource for capital. We acknowledge that the use of extraction is an inadequate metaphor, and that such practices produce value.

Laboratory A place of labor, but where labor is subordinated to the task of elaboration. In the lab, there are consistent procedures, forms of regularity but not necessarily to produce a standard commodity, in the way that a factory does. The lab produces differences that can be tested, verified, stabilized, and that can become the prototypes for new forms of organization of some aspect of the world, be it the material or human world. The laboratory is not really always about the production of knowledge, the generation of new regularities that will be more efficient, more rational, more fictionless. Sometimes the lab seems to exist for no other reason than the desire to experiment on precarious bodies, to no end, and for no reason. The lab has no necessary relation to reason, to enact power simply as power. (See Chapter 1: Precarity Lab.)

Overcommons We coined this term to describe technocratic societies characterized by radical inequality, a class of people whose richly compensated labor is enabled by the precarious labor of others who perform both the "gig" work that results in cheap and convenient car rides, places to rent during vacations, and task labor, and those who do undergig work. This technocratic overcommons produces the "undercommons" (Stefano Harney and Fred Moten, *The Undercommons: Fugitive Planning & Black Study* (Brooklyn, NY: Autonomedia, 2013)). (See Chapter 2: The Undergig.)

Precarity Precarity is a state of being and lived experience of insecurity, loss of control, and unpredictability of one's world. Precarity's etymology is derived from Latin *precarius*: given as a favor, at the pleasure of another person (*OED*). Our work highlights the differential circumstances in which precarity arises, extracting resources from and implicating racial, ethnic, and sexual minorities, women, indigenous people, and migrants who occupy extractive and depleted zones.

Social safety networks Systems of collective resource-sharing and emotional support, often across geographical distance, that emerge under conditions of decreasing state support.

Spirals/the screw of exploitation We use the analogy and the figure of the spirals as a way to highlight the recursive, circular, and penetrative nature of precarity in relation to capitalism and techno governance. One form of precarity, say, housing insecurity, disability, or educational debt, results in other forms of precarity – leading to an enmeshment and cycling of insecurity that is almost impossible to escape from. (See Chapter 2: The Undergig; Chapter 4: The Widening Gyre of Precarity.)

Toxicity The spread of environmental harm and vulnerability in the depletion economy. Digital industries create toxic matter that spreads and latches onto all kinds of bodies, some more than others. It is both a cultural condition and a material state of being.

Undergig Cheap labor is a precondition of the gig economy, hence we call these workers the undergig. Undergig workers perform the often invisible labor needed to create the conditions of digital life for everyone else. Electronics production that extracts value from depleted zones and factory workers, and produces toxicity, often overlaps with the "global south" category but also exceeds it. Being female, poor, offshored, poor, and non-white greatly increases your chances of being an undergig worker. The undergig is post-territorial: white men in Europe and the US can find themselves in this "sunken place." (see Chapter 8: The Affronted Class; see also Chapter 2: The Undergig.)

Bibliography

Adair, Cassius. "White Trans Politics and the Early Internet." Paper Presentation, American Studies Association Annual Meeting, Atlanta, Georgia, November 8–11, 2018.

Aizura, Aren Z. *Mobile Subjects: Transnational Imaginaries of Gender Reassignment.* Durham, NC: Duke University Press, 2018.

Allied Media Projects. *Our Data Bodies: Digital Defense Playbook.* Detroit, MI: Allied Media Projects, 2019.

Anderson, Chris. *Makers: The New Industrial Revolution.* New York: Crown Business, 2012.

Anzaldúa, Gloria. "Foreword." In *This Bridge Called My Back: Writings by Radical Women of Color,* edited by Cherríe Morraga and Gloria Anzaldúa, iii–iv. New York: Kitchen Table: Women of Color Press, 1983.

Apter, David E., Herbert J. Gans, Ruth Horowitz, Gerald D. Jaynes, William Kornblum, James F. Short, Gerald D. Suttles, and Robert E. Washington. "The Chicago School and the Roots of Urban Ethnography: An Intergenerational Conversation with Gerald D. Jaynes, David E. Apter, Herbert J. Gans, William Kornblum, Ruth Horowitz, James F. Short Jr., Gerald D. Suttles, and Robert E. Washington." *Ethnography* 10, no. 4 (2009): 375–396.

Baucom, Ian. *Specters of the Atlantic: Finance, Capital, Slavery, and the Philosophy of History.* Durham, NC: Duke University Press, 2005.

Baumkel, Max. "The Invisible Presence of Trans-Bodies: Unpacking Regimes of Visibility and Visuality Through Tom Cho's *Look Who's Morphing.*" Master's Thesis, Vanderbilt University, 2015.

Beckert, Steven. *Empire of Cotton: A Global History.* New York: Vintage Press, 2014.

Beckett, Samuel. *Samuel Beckett's Molloy, Malone Dies, The Unnamable,* edited by Harold Bloom. New York: Chelsea House Publishers, 1998.

Benjamin, Ruha. *Race After Technology: Abolitionist Tools for the New Jim Code*. Medford, MA: Polity, 2019.

Berardi, Franco Bifo. *The Soul at Work: From Alienation to Autonomy*, translated by Francesca Cadel and Giuseppina Mecchia. Los Angeles, CA: Semiotext(e), 2009.

Berlant, Lauren. *Cruel Optimism*. Durham, NC: Duke University Press, 2011.

Bichell, Rae Ellen. "Scientists Start to Tease out the Subtler Ways Racism Hurts Health." *NPR*, November 11, 2017. www.npr.org/sections/health-shots/2017/11/11/562623815/scientists-start-to-tease-out-the-subtler-ways-racism-hurts-health.

Bowker, Geof. "How to be Universal: Some Cybernetic Strategies, 1943–1970." *Social Studies of Science* 23, no. 1 (1993): 107–127.

Brown, H. Claire. "How an Algorithm Kicks Small Businesses Out of the Food Stamps Program on Dubious Fraud Charges." *The Counter*, October 8, 2018. https://thecounter.org/usda-algorithm-food-stamp-snap-fraud-small-businesses/.

Browne, Simone. *Dark Matters: On the Surveillance of Blackness*. Durham, NC: Duke University Press, 2015.

Butler, Judith. *Undoing Gender*. New York: Routledge, 2004.

Button, John. *A Dictionary of Green Ideas: Vocabulary for a Sane and Sustainable Future*. New York: Routledge, 1998.

Byrd, Jodi. *Transit of Empire: Indigenous Critiques of Colonialism*. Minneapolis, MN: University of Minnesota Press, 2011.

Casidy, Ciaran and Adrian Chen. "The Moderators." *Field of Vision*, April 2017. https://vimeo.com/213152344.

Chaar-López, Iván. "Sensing Intruders: Race and the Automation of Border Patrol." *American Quarterly* 71, no. 2 (2019): 495–518.

Chakrabarty, Dipesh. *Provincializing Europe: Postcolonial Thought and Historical Difference*. Princeton, NJ: Princeton University Press, 2000.

Chen, Adrian. "The Laborers Who Keep Dick Pics and Beheadings Out of Your Facebook Feed." *Wired*, October 23, 2014. www.wired.com/2014/10/content-moderation/.

Chen, Mel Y. "Toxic Animacies, Inanimate Affections." *GLQ: A Journal of Lesbian and Gay Studies* 17, nos 2–3 (2011): 265–286.

Chen, Mel Y. "Unpacking Intoxication, Racialising Disability." *Medical Humanities* 41, no. 1 (2015): 25–29.

Chun, Wendy Hui Kyong. *Control and Freedom: Power and Paranoia in the Age of Fiber Optics*. Cambridge, MA: The MIT Press, 2008.

Clare, Eli. *Brilliant Imperfection: Grappling with Cure*. Durham, NC: Duke University Press, 2017.

Clark, Krissy. "The Disconnected," *Slate*, January 3, 2016. https://slate.com/news-and-politics/2016/06/welfare-to-work-resulted-in-neither-welfare-nor-work-for-many-americans.html.

Davis, Angela Y. "Reflections on the Black Woman's Role in the Community of Slaves." *The Black Scholar: Journal of Black Studies and Research* 4 (1971): 2–15.

Davis, Angela Y. *Freedom is a Constant Struggle: Ferguson, Palestine, and the Foundations of a Movement*. Chicago, IL: Haymarket Books, 2016.

Deegan, Mary Jo. *Race, Hull-House, And The University Of Chicago: A New Conscience Against Ancient Evils*. Westport, CT: Praeger, 2002.

Deleuze, Gilles. "Postscript on the Societies of Control." *October* 59 (1992): 3–7.

Dyck, Darryl. "Trans Mountain, Trudeau and First Nations." *The Globe and Mail*, April 27, 2018. www.theglobeandmail.com/politics/article-trans-mountain-kinder-morgan-pipeline-bc-alberta-explainer/.

Escobar, Arturo. *Encountering Development: The Making and Unmaking of the Third World*. Princeton, NJ: Princeton University Press, 1994.

Eubanks, Virginia. *Automating Inequality: How High-Tech Tools Profile, Police, and Punish the Poor*. New York: St. Martin's Press, 2017.

Fisher, Anna Watkins. "User Be Used: Leveraging the Play in the System." *Discourse* 36, no. 3 (2014): 383–399.

Forbes, Rosie. "What 'Detective Pikachu' Got Wrong About Disability." *The Mighty*, May 17, 2019. https://themighty.com/2019/05/detective-pikachu-disability-villain/?utm_source=pin_board_disability&utm_medium=pinterest&utm_campaign=pin_disability_2019week18.

Foucault, Michel. *The Order of Things*. New York: Routledge, 2002.

Franklin, Seb. *Control: Digitality as Cultural Logic*. Cambridge, MA: The MIT Press, 2015.

Friedman, Zach. "99% Rejected for Student Loan Forgiveness – Again." *Forbes*, September 9, 2019. www.forbes.com/sites/zackfriedman/2019/09/09/99-rejected-for-student-loan-forgiveness-again/#69a05d1d6675.

Garfield, Rachel, Robin Rudowitz, and Kendal Orgera. "Understanding the Intersection of Medicaid and Work: What Does the Data Say?" *Kaiser Family Foundation*, August 8, 2019. www.kff.org/medicaid/issue-brief/understanding-the-intersection-of-medicaid-and-work-what-does-the-data-say/.

Gault, Matthew. "Hunting Silicon Valley's Doomsday Bunkers." *Vice*, May 20, 2019. www.vice.com/en_us/article/kzmzyx/hunting-silicon-valleys-doomsday-bunkers-in-new-zealand-documentary.

Georgakas, Dan and Marvin Surkin. *Detroit, I Do Mind Dying: A Study in Urban Revolution*. New York: Southend Press, 1998.

Giddens, Anthony. *The Nation-State and Violence: Volume Two of A Contemporary Critique of Historical Materialism*. Cambridge, MA: Polity, 1985.

Gill-Peterson, Jules. "The Techical Capacities of the Body: Assembling Race, Technology, and Transgender." *Transgender Studies Quarterly* 1, no. 3 (2014): 402–418.

Gilmore, Ruth Wilson. *Golden Gulag: Prisons, Surplus, Crisis, and Opposition in Globalizing California*. Berkeley, CA: University of California Press, 2007.

Gray, Mary L. and Siddharth Suri. *Ghost Work: How to Stop Silicon Valley from Building a New Global Underclass*. Boston, MA: Houghton Mifflin Harcourt, 2019.

Guiley, Rosemary. *The Encyclopedia of Witches, Witchcraft and Wicca*. New York: Facts On File, 2008.

Haraway, Donna. "Situated Knowledges: The Science Question in Feminism and the Privilege of Partial Perspective." *Feminist Studies* 14, no. 3 (1988): 575–599.

Haraway, Donna. *Modest_Witness@Second_Millennium.FemaleMan©_Meets_OncoMouse™: Feminism and Technoscience*. New York: Routledge, 1997.

Harney, Stefano and Fred Moten. *The Undercommons: Fugitive Planning & Black Study*. Brooklyn, NY: Autonomedia, 2013.

Hartman, Saidiya. "The Anarchy of Colored Girls Assembled in a Riotous Manner." *South Atlantic Quarterly* 117, no. 3 (2018): 465–490.

Hartman, Saidiya. *Wayward Lives, Beautiful Experiments: Intimate Histories of Social Upheaval*. New York: W.W. Norton & Company, Inc., 2019.

Hazboun, Areej. "New Apps Help Palestinians Navigate Israeli Checkpoints." *AP News*, November 18, 2015. https://apnews.com/117d524e2aba4ab19ec01c3d3e50aa06.

Hersman, Erik. "Reporting Crisis via Texting." Filmed November 2009. TED video, 3:44. www.ted.com/talks/erik_hersman_reporting_crisis_via_texting/details?language=bi.

Hill, Kashmir. "The Secretive Company That Might End Privacy as We Know It." *The New York Times*, January 18, 2020. www.nytimes.com/2020/01/18/technology/clearview-privacy-facial-recognition.html.

Hochschild, Arlie Russel. "Emotion Work, Feeling Rules, and Social Structure." *American Journal of Sociology* 85, no. 3 (1979): 551–575.

Hollister, Sean. "Google Contractors Reportedly Targeted Homeless People for Pixel 4 Facial Recognition." *The Verge*, October 2, 2019. www.theverge.com/2019/10/2/20896181/google-contractor-reportedly-targeted-homeless-people-for-pixel-4-facial-recognition.

Horwitz, Rainey. "The Jane Collective (1969–1973)." *The Embryo Project Encyclopedia*, August 7, 2017. https://embryo.asu.edu/pages/jane- collective-1969-1973.

Houser, Micah. "The High Price of Freedom for Migrants in Detention." *The New Yorker*, March 12, 2019. www.newyorker.com/news/news-desk/the-high-price-of-freedom-for-migrants-in-detention.

Howe, LeAnn. "The Chaos of Angels." *Callaloo* 17, no. 1 (1994): 108–114.

Idyll Dandy Arts. "About Ida." https://idylldandyarts.tumblr.com/about.

Irani, Lily. "The Cultural Work of Microwork." *New Media & Society* 17, no. 5 (2015): 720–739.

Jewish Voices for Peace. “Deadly Exchange: The Dangerous Consequences of American Law Enforcement Trainings in Israel.” Deadly Exchange, September 2018. https://deadlyexchange.org/wp-content/uploads/2019/07/Deadly-Exchange-Report.pdf.

Johnson, Walter. *River of Dark Dreams: Slavery and Empire in the Cotton Kingdom*. Cambridge, MA: Harvard University Press, 2017.

Joque, Justin. *Infidel Mathematics.* New York: Verso, forthcoming.

Kamil, Meryem. “Towards Decolonial Futures: New Media, Digital Infrastructures, and Imagined Geographies of Palestine.” PhD Dissertation, University of Michigan, 2019.

Kamil, Meryem. “Post Spatial, Post Colonial: Accessing Palestine in the Digital.” *Social Text*, 38, no. 3 (144) (2020): 55–82.

Keyes, Os, Nikki Stevens, and Jacqueline Wernimont. “The Government is Using the Most Vulnerable People to Test Facial Recognition Software.” *Slate*, March 17, 2019. https://slate.com/technology/2019/03/facial-recognition-nist-verification-testing-data-sets-children-immigrants-consent.html.

Knorr Cetina, Karin. *Epistemic Cultures: How the Sciences Make Knowledge.* Cambridge, MA: Harvard University Press, 1999.

Lancianese, Adelina. “Before Black Lung, the Hawk Nest Disaster Killed Hundreds.” *NPR*, January 20, 2019. www.npr.org/2019/01/20/685821214/before-black-lung-the-hawks-nest-tunnel-disaster-killed-hundreds.

Latour, Bruno and Steve Woolgar. *Laboratory Life: The Construction of Scientific Facts*, 2nd edn. Princeton, NJ: Princeton University Press, 1986.

Library of Congress. “Fugitive Slave Ads: Topics in Chronicling America.” Library of Congress Research Guides. https://guides.loc.gov/chronicling-america-fugitive-slave-ads.

Lindtner, Silvia. *Prototype Nation: China and the Contested Promise of Innovation*. Princeton, NJ: Princeton University Press, 2020.

Lynskey, Dorian. “How Dangerous is Jordan B. Peterson, The Rightwing Professor Who ‘Hit a Hornets’ Nest?’ ” *The Guardian*, February 7, 2018. www.theguardian.com/science/2018/feb/07/how-dangerous-is-jordan-b-peterson-the-rightwing-professor-who-hit-a-hornets-nest.

Marx, Karl. *Capital, Volume I: A Critique of Political Economy*, translated by Ben Fowkes. New York: Penguin, 1976.

Mauss, Marcel. *The Gift: Forms and Functions of Exchange in Archaic Societies*, translated by W.D. Halls. New York: Norton, 1967.

Mbembe, Achille. *Critique of Black Reason*, translated by Laurent Dubois. Durham, NC: Duke University Press, 2017.

McCoy, Alfred W. *Policing America's Empire: The United States, the Philippines, and the Rise of the Surveillance State*. Madison, WI: The University of Wisconsin Press, 2009.

Mezzadra, Sandro and Brett Neilson. *The Politics of Operations: Excavating Contemporary Capitalism*. Durham, NC: Duke University Press, 2019.

Miller, Michael E. "This Company is Making Millions from America's Broken Immigration System." *The Washington Post*, March 9, 2017. www.washingtonpost.com/local/this-company-is-making-millions-from-americas-broken-immigration-system/2017/03/08/43abce9e-f881-11e6-be05-1a3817ac21a5_story.html.

Muñoz, José Esteban. "Feeling Brown, Feeling Down: Latina Affect, the Performativity of Race, and the Depressive Position." *Signs* 31, no. 3 (2006): 675–688.

Murphy, Michelle. *The Economization of Life*. Durham, NC: Duke University Press, 2017.

Negri, Antonio. *Marx Beyond Marx: Lessons on the Grundrisse*, translated by Harry Cleaver. Brooklyn, NY: Autonomedia, 1992.

Negri, Antonio. *The Politics of Subversion: A Manifesto for the Twenty-First Century*. Malden, MA: Polity, 2005.

Nixon, Rob. *Slow Violence and the Environmentalism of the Poor*. Cambridge, MA: Harvard University Press, 2011.

Ong, Aihwa. *Neoliberalism as Exception: Mutations in Citizenship and Sovereignty*. Durham, NC: Duke University Press, 2006.

Patel, Raj and Jason W. Moore. *A History of the World in Seven Cheap Things: A Guide to Capitalism, Nature, and the Future of the Planet*. Oakland, CA: University of California Press, 2017.

Peters, Torrey. *Infect Your Friends and Loved Ones.* CreateSpace Independent Publishing Platform, 2016.

Power-Sotomayor, Jade Y. "From Soberao to Stage: Afro-Puerto Rican and the Speaking Body." In *The Oxford Handbook of Dance and Theater,* edited by Nadine George-Graves. New York: Oxford University Press, 2015.

Precarity Lab. "Digital Precarity Manifesto." *Social Text* 37, no. 4 (2019): 77–93.

Préciado, Paul B. *Testo Junkie: Sex, Drugs, and Biopolitics in the Pharmacopornographic Era.* New York: The Feminist Press, 2013.

Proctor, Jennifer. "A Failure of Imagination: The Role of Disability in Avatar." *Media Commons,* August 10, 2010. http://mediacommons.org/imr/2010/08/03/failure-imagination-role-disability-avatar.

Robinson, Cedric. *Black Marxism: The Making of the Black Radical Tradition.* Durham, NC: Duke University Press, 1993.

Rose, Sarah F. *No Right to Be Idle: The Invention of Disability, 1840s–1930s.* Chapel Hill, NC: University of North Carolina Press, 2017.

Ross, Andrew. *Bird on Fire: Lessons from the World's Least Sustainable City.* New York: Oxford University Press, 2011.

Rukeyser, Muriel. *The Book of the Dead.* Morgantown, WV: University of West Virginia Press, 2018.

Saxton, Alexander. *The Indispensable Enemy: Labor and the Anti-Chinese Movement in California.* Berkeley, CA: University of California Press, 1971.

Scholz, Trebor, ed. *Digital Labor: The Internet as Playground and Factory.* New York: Routledge, 2013.

Shapin, Steven and Simon Schaffer. *Leviathan and the Air-Pump: Hobbes, Boyle, and the Experimental Life.* Princeton, NJ: Princeton University Press, 2018.

Sharpe, Christina. *In the Wake: On Blackness and Being.* Durham, NC: Duke University Press, 2016.

Shields, L.M., W.H. Wiese, B.J. Skipper, B. Charley, and L. Benally. "Navajo Birth Outcomes in the Shiprock Uranium Mining Area." *Health Physics* 63, no. 5 (1992): 542–551.

Shotwell, Alexis. *Against Purity: Living Ethically in Compromised Times.* Minneapolis, MN: University of Minnesota Press, 2016.

Simpson, Cam. "American Chipmakers Had a Toxic Problem. Then They Outsourced It." *Bloomberg Businessweek,* June 15, 2017. www.bloomberg.com/news/features/2017-06-15/american-chipmakers-had-a-toxic-problem-so-they-outsourced-it.

Smyth, Araby, Jess Linz, and Lauren Hudson. "A Feminist Coven in the University." *Gender, Place and Culture: A Journal of Feminist Geography* (2019): 2.

Srnicek, Nick. *Platform Capitalism.* Malden, MA: Polity, 2017.

Standing, Guy. *The Precariat: The New Dangerous Class.* New York: Bloomsbury, 2011.

Steele, Catherine Knight. "Signifyin' Bitching and Blogging: Black Women and Resistance Discourse Online." In *The Intersectional Internet: Race, Sex, Class and Culture Online,* edited by Safiya Umoja Noble and Brendesha M. Tynes, 73–93. New York: Peter Lang, 2016.

Stoler, Ann Laura. *Duress: Imperial Durability in Our Times.* Durham, NC: Duke University Press, 2016.

Sue Ramírez, Catherine. "Deus Ex Machina: Tradition, Technology, and the Chicanafuturist Art of Marion C. Martinez." *Aztlán* 29, no. 2 (2004): 55–92.

Tawil-Souri, Helga and Miriyam Aouragh. "Intifada 3.0? Cyber Colonialism and Palestinian Resistance." *The Arab Studies Journal* 22, no. 1 (2014): 102–133.

Thakor, Mitali. "Digital Apprehensions: Policing, Child Pornography, and the Algorithmic Management of Innocence." *Catalyst: Feminism, Theory, Technoscience* 4, no. 1: 1–16.

Thompson, E.P. "Time, Work Discipline, and Industrial Capitalism." *Past & Present* 38 (1967): 56–97.

Towghi, Fouzieyha and Kalindi Vora. "Bodies, Markets, and Experiments in South Asia." *Ethnos: Journal of Anthropology* 70, no. 1 (2014): 1–18.

Tronti, Mario. "Factory and Society." *Operaismo in English,* June 13, 2013. https://operaismoinenglish.wordpress.com/2013/06/13/factory-and-society/.

Tronti, Mario. *Workers and Capital,* translated by David Broder. New York: Verso, 2019.

Tsing, Anna Lowenhaupt. *The Mushroom at the End of the World: On the Possibility of Life in Capitalist Ruins.* Princeton, NJ: Princeton University Press, 2015.

Vora, Kalindi. *Life Support: Biocapital and the New History of Outsourced Labor.* Minneapolis, MN: University of Minnesota Press, 2015.

Wajcman, Judy. *Feminism Confronts Technology.* University Park, PA: Pennsylvania State University Press, 1991.

Walker, Brett. *Toxic Archipelago: A History of Industrial Disease in Japan.* Seattle, WA: University of Washington Press, 2010.

Wark, McKenzie. *A Hacker Manifesto.* Cambridge, MA: Harvard University Press, 2004.

Warren, Elizabeth and Amelia Warren Tyagi. *The Two-Income Trap: Why Middle-Class Parents are Going Broke.* New York: Basic Books, 2003.

Weber, Max. *Economy and Society: A New Translation,* edited and translated by Keith Tribe. Cambridge, MA: Harvard University Press, 2019.

Wynn, Natalie. *ContraPoints.* www.contrapoints.com.

Wynter, Sylvia. "Unsettling the Coloniality of Being/Power/Truth/Freedom: Towards the Human, After Man its Overrepresentation – An Argument." *CR: The New Centennial Review* 3, no. 3 (2003): 257–337.

Yeats, William Butler. "The Second Coming." In *The Collected Works of W.B. Yeats,* 2nd edn, edited by Richard J. Finneran, 187. New York: Scribner Paperback Poetry, 1996.

Zuboff, Shoshana. *The Age of Surveillance Capitalism: The Fight for a Human Future at the New Frontier of Power.* New York: Public Affairs, 2019.

Zuckerberg, Mark. "Opening Statement to the Senate Judiciary and Commerce Committees on Facebook Data Privacy." *American Rhetoric,* April 10, 2018. www.americanrhetoric.com/speeches/markzuckerbergcongressopeningstmt.htm.

Index